THE NORTH AMERICAN WHISTLING-DUCKS, POCHARDS, AND STIFFTAILS

The North American Whistling-Ducks, Pochards, and Stifftails

Paul A. Johnsgard

School of Biological Sciences
University of Nebraska–Lincoln

Zea Books, Lincoln, Nebraska 2017

*My five volumes (Johnsgard, 2016–17) on North American waterfowl
are dedicated to the memory of Hans Albert ("Al") Hochbaum (1911–1988),
who proved that waterfowl monographs can be written by one with the eyes of a naturalist,
the pen of an artist, and the heart of a poet.*

ISBN: 978-1-60962-110-0

https://doi.org/10.13014/K28913SK

Composed in Adobe Garamond and Imprint MT Shadow types.

Spot illustrations: p. 1, Lesser scaup, male; p. 2, Redhead, pair; p. 6, Ring-necked duck, pair;
p. 20, Black-bellied whistling-duck, pair; p. 22, Black-bellied whistling-duck, adult;
p. 52, Canvasback, pair; p. 115, Greater scaup, male;
p. 129, Ruddy duck, male-tail-cocking in front of female;
p. 183, Ruddy duck, courting group

Zea Books are published by the University of Nebraska–Lincoln Libraries
Electronic (pdf) edition available online at http://digitalcommons.unl.edu/zeabook/
Print edition available from http://www.lulu.com/spotlight/unllib

UNIVERSITY OF
Nebraska
Lincoln®

Abstract

Although the 12 species representing three waterfowl tribes described in this volume are not closely related, they fortuitously provide an instructive example of adaptive evolutionary radiation within the much larger waterfowl lineage (the family Anatidae), especially as to their divergent morphologies, life histories, and social behaviors.

The whistling-ducks (*Dendrocygna*), with three known North American species, are notable for their permanent pair-bonds, extended biparental family care, and strong social cohesion. In contrast, males of the five typical pochards of North American diving ducks (*Aythya*) establish monogamous pair-bonds that are maintained only long enough to assure that the female's eggs are fertilized. The endpoint of this behavioral gradient, promiscuity or polygyny, exists among at least some of the typical stifftails (*Oxyura*). Such diverse reproductive strategies have exerted powerful evolutionary influences on interspecies variations in sexual dimorphism, sexual behavior, anatomy, ecology, and other traits.

For example, the locomotory anatomies and swimming abilities of the three groups vary greatly. The whistling-ducks are all long-legged species that are well adapted to both walking and swimming but are poor at diving. All five North American pochards swim and dive superbly but walk awkwardly on land. They usually nest above water in emergent vegetation or very close to water. The two North American stiff-tailed ducks are among the most accomplished swimmers and divers of all waterfowl. They construct easily accessible, semifloating reed nests among emergent vegetation, but their hind-set legs and large feet have become so specialized for aquatic locomotion that they can barely walk on land.

The flying abilities of the three groups also diverge greatly. The pochards are all remarkably swift fliers: the canvasback has been clocked in excess of 70 mph, and radio-tagged lesser scaups have flown more than 2,000 miles from North Dakota to Cuba in as few as three days. In contrast, the whistling-ducks are among the slowest flyers of all North American ducks. Their tropical populations are mostly nonmigratory, although fulvous whistling-ducks have been known to cross the 90-mile ocean gap between Florida and Cuba. Lastly, stiff-tailed ducks have such small wing areas and correspondingly high wing-loading that they typically dive, rather than fly, to escape danger. Ruddy ducks can take flight only after a running start, reaching air speeds of only about 45 miles per hour. Their seasonal migrations are relatively prolonged, so they are among the last waterfowl to arrive on their northernmost breeding areas. Such long migrations are probably undertaken by spreading them out over many shorter segments.

This volume includes more than 63,000 words, plus some 200 maps, photos, drawings, and sketches, and nearly 650 literature citations. It is the last of five volumes that describe all 55 waterfowl species that have been historically documented in North America; collectively, the volumes total over 300,000 words, with nearly 3,000 literature citations, and more than 600 maps, photos, drawings, and sketches.

Contents

Distribution Maps

Figures

Photographs

Preface

This volume includes ten species of ducks that collectively represent three relatively disparate tribes within the waterfowl family Anatidae. One of the three groups, the whistling-ducks, is much more closely related to geese and swans than it is to typical ducks. A second assemblage includes five species that are efficient diving birds; colloquially known as bay ducks, inland diving ducks, or pochards, they are dispersed across nearly all continents. A third group, the stiff-tailed ducks, is only distantly related to these and other waterfowl, and its members are highly adapted for diving and underwater swimming, but in the process have sacrificed aerial speed and maneuverability.

Beyond these species' great importance to professional wildlife biologists and conservationists, they generate high economic benefits that result from sport hunting expenditures; over 100,000 individuals of these ducks are shot annually in the United States plus thousands more in Canada and Mexico. They also provide many nonconsumptive benefits, such as recreational birding, wildlife photography, and general nature appreciation. Finally, because of their beauty, widespread occurrence, fascinating behavior, and relative conspicuousness, they foster a broad popular interest in water birds, wetland values, and general environmental conservation.

For biologists, these species illustrate a remarkable diversity in their ecologies, social-sexual behaviors, and evolutionary adaptations that spans almost the entire range of these characteristics within the family Anatidae. Their pairing systems range from permanent monogamy (whistling-ducks) to weak or early absent pair-bonds (stifftails), their diving abilities vary from limited diving and underwater capabilities (whistling-ducks) to being superb divers (pochards), and from being the poorest flyers of all Anatidae (whistling-ducks) to having sometimes transhemispheric migratory movements (pochards). All together, these three waterfowl groups provide a fascinating insight into the even greater range of evolutionary adaptions to be found among the roughly 150 extant species present across the entire family Anatidae.

This contribution is the last of five volumes that collectively update my *Waterfowl of North America* (Johnsgard, 1975). Together with four earlier volumes (Johnsgard, 2016a, 2016b, 2016c, 2017a), it completes a survey of every species of North America's breeding and nonbreeding waterfowl, and provides updated biological information on the 55 species that I first documented over four decades ago. Collectively, the five volumes contain more than 300,000 words, 53 maps, 111 drawings, 130 photos and more than 200 behavioral or anatomical sketches. There are also more than 3,000 literature citations; about one tenth of which are duplicated in two or more volumes. The copyright in all the text, photographs, maps, and drawings belongs to me.

As with my previous monographs already placed into the UNL DigitalCommons library, I owe a huge debt of gratitude to Paul Royster, Coordinator of Scholarly Communications for the University of Nebraska–Lincoln Libraries and publisher of Zea Books, for accepting and carrying this large project through to completion, and for achieving such a satisfying result. I also thank Linnea Fredrickson for her tireless editing and unflagging interest in this long project.

Paul A. Johnsgard
Foundation Regents Professor Emeritus
Biological Sciences
University of Nebraska–Lincoln

I. Introduction

Reviewing the recent literature on the whistling-ducks, pochards, and stiff-tailed ducks has been a delightful task, as the stiff-tailed ducks have been a passionate interest of mine since childhood. I saw my first live whistling-ducks only after completing my graduate work. I immediately became very fond of them, and even once imagined I would someday write a comprehensive book on whistling-ducks. Other writing interests intervened, including writing a book on the stiff-tailed ducks. Regrettably there is still not a single full-length book on the biology of whistling-ducks, nor indeed one on the equally interesting group of pochards. However, H. A. Hochbaum's *Canvasback on a Prairie Marsh* provided a wonderful glimpse into the biology of one of the pochard species and, while I was an undergraduate student, evoked in me a vision of someday also writing a waterfowl monograph.

Grouping the whistling-ducks, pochards, and stiff-tailed ducks into a single volume might seem taxonomically counterintuitive, but the very diversity of these three groups invites thoughtful comparisons. Basically, all of them share the same fundamental waterfowl anatomy (Fig. 1), with major variations in the length and locations of their legs and the size of their feet, and especially the area of foot webbing relative to their body mass. Thus, the black-bellied whistling-duck, which rarely dives but often perches, has relatively small feet but very long hind toes. It has an adult body mass averaging about 800 grams and a tarsal length of about 60 mm.

As a rough comparison, the canvasback has a body mass of about 1,200 grams, a wing length of about 240 mm, and a tarsal length of about 40 mm. The ruddy duck has a body mass of about 500 grams, a wing length of about 150 mm, and a tarsal length of about 30 mm. The notably short tarsi of canvasbacks and ruddy ducks relative to their body mass results in much shorter paddling strokes than is true of whistling-ducks, but their relatively longer toes and larger web surface areas of the two former species probably compensate for their shorter tarsi. Likewise the long wing-to-tarsal-length ratio of the canvasback (about 5:1) compared to the shorter ratios of the whistling-duck and ruddy duck (about 3.3:1) reflect the much greater flying efficiency of the canvasback over the other two species.

In another respect the ruddy duck (and probably all other *Oxyura* species) is remarkable, and that is in the male's tracheal air sac (Fig. 1H). This balloon-like structure is connected by an air duct to the trachea and, when inflated during sexual display, substantially increases the size of the neck (Roberts, 1932, p. 274). Another remarkable trait that occurs in males of several and probably all species of *Oxyura* is their remarkably long, coiled copulatory structures (Fig. 1I). In the Argentine lake duck (*O. vittata*), this reaches a length of up to eight inches when extended, equal to about half the bird's total body length. (McCracken, 2000; Cocker et al., 2002). Similar copulatory structures are known to occur in the ruddy duck and the Australian blue-billed duck (*O. australis*). Although all species of waterfowl have extrusive copulatory organs that are probably related to the difficulties of achieving effective aquatic fertilization, those of the highly aquatic stifftails are the largest and most elaborate in the entire waterfowl family.

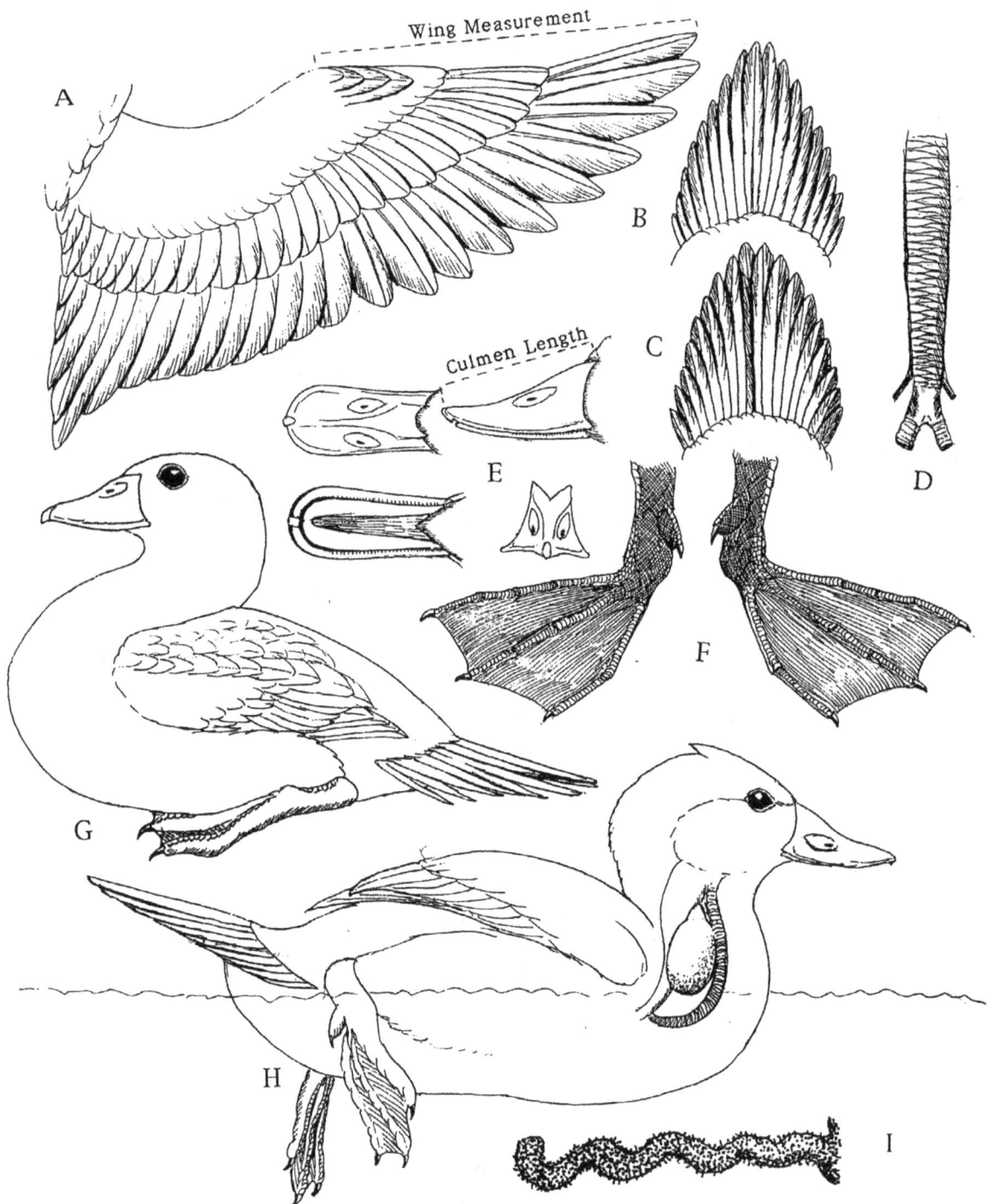

Fig. 1. External topography and anatomy of typical stiff-tailed ducks (*Oxyura*), including (A) upper wing feathers, (B–C) dorsal and ventral views of tail, (D) trachea, (E) bill anatomy, (F) foot anatomy, (G) resting profile of masked duck, and (H) swimming profile of ruddy duck, including male's tracheal air sac, and (I) the extended phallus of *O. vittata* (after McCracken, 2000).

Among these three distinctive taxonomic groups, there is also much foraging and food-related diversity. All the whistling-ducks are strongly vegetarian, as are the typical pochards (such as the canvasback and redhead), whereas other pochards (such as the scaups) are adapted for consuming invertebrate foods. Most of the typical stiff-tailed ducks (*Oxyura, Biziura*) also feed largely on benthic invertebrates.

Depending on their foods and foraging methods, these North American duck species have noticeable differences in their bill shape and morphology across the three tribes (Fig. 2). Associated with their anatomical, locomotory and foraging diversity are a wide variety of foraging adaptations. For example, the fulvous whistling-duck (Fig. 2A) is primarily a strainer of aquatic vegetation, unlike the black-bellied whistling-duck (Fig. 2B), which is typically a terrestrial grazer, and there are marked differences in the bill morphology of these two species (Bolen and Rylander, 1974b).

Among the pochards and stiff-tailed ducks there is a considerable variation in bill form and functional structure (Kear, 2005). At one extreme is the long, narrow, and deep bill of the canvasback (Fig. 2C–D), which is well adapted to grasping and extracting vegetation from the bottoms of ponds and probably evolved for probing and grasping tubers (Tome and Wrubleski, 1988). The canvasback's bill form grades into the wider and flatter bill of the scaups (Fig. 2I–2L), with the redhead (Fig. 2E–F) and ring-necked duck (Fig. 2G–H) being intermediate. The ring-necked bill form is probably better adapted to grasping vegetation and straining seeds, whereas the scaups' bills are efficient at tactilely detecting and grasping invertebrate prey (Tome and Wrubleski, 1988).

The ruddy duck, a typical *Oxyura* stifftail, has a bill (Fig. 2N) that is highly adapted to the tactile detection and straining of tiny food materials from muddy bottom sediments (Lagerquist and Ankney, 1989). The bill of the masked duck (Fig. 2M) is unusually sturdy by comparison with the *Oxyura* type and instead resembles those of pygmy geese (*Nettapus* spp.), which consume the seeds of water-lilies and other aquatic plants. The masked duck's bill might be similarly adapted for grasping and consuming seeds, roots, and stems of aquatic plants (Johnsgard and Carbonell, 1996), although this species' diet is still inadequately known.

A good deal of time is spent by all waterfowl in self-maintenance behavior and associated care of their integument, especially their plumage (McKinney, 1953). This is achieved by obtaining oil from their preen gland (uropygial gland) at the dorsal base of the tail (Fig. 3A) and spreading it over their plumage while preening (Fig. 3A–3C).

Other actions, if not so obviously self-maintenance in function, are various forms of "comfort" activities, such as bill-stretching ("yawning") (Fig. 3D), wing-and-leg stretching (Fig. 3E), two-wing stretching (Fig. 3F), wing-flapping (Fig. 3G), rotary head-shaking (Fig. 3H), and a "general shake" of the body (Fig. 3J). While bathing energetically, waterfowl often perform somersaulting (Fig. 3I), apparently to wet their plumage more thoroughly. They sometimes use head-flicking to remove water or solid particles from their bill (Fig. 3K), and perform repeated head-dipping and head-retraction movements during bathing (Fig. 3M).

Over time, many of these postures and movements have evolved and become stereotyped ("ritualized") to also serve in social situations as important specific signals, or "displays." Thus, head-dipping is a common precopulatory display in whistling-ducks, preening-behind-the-wing serves a similar function in pochards,

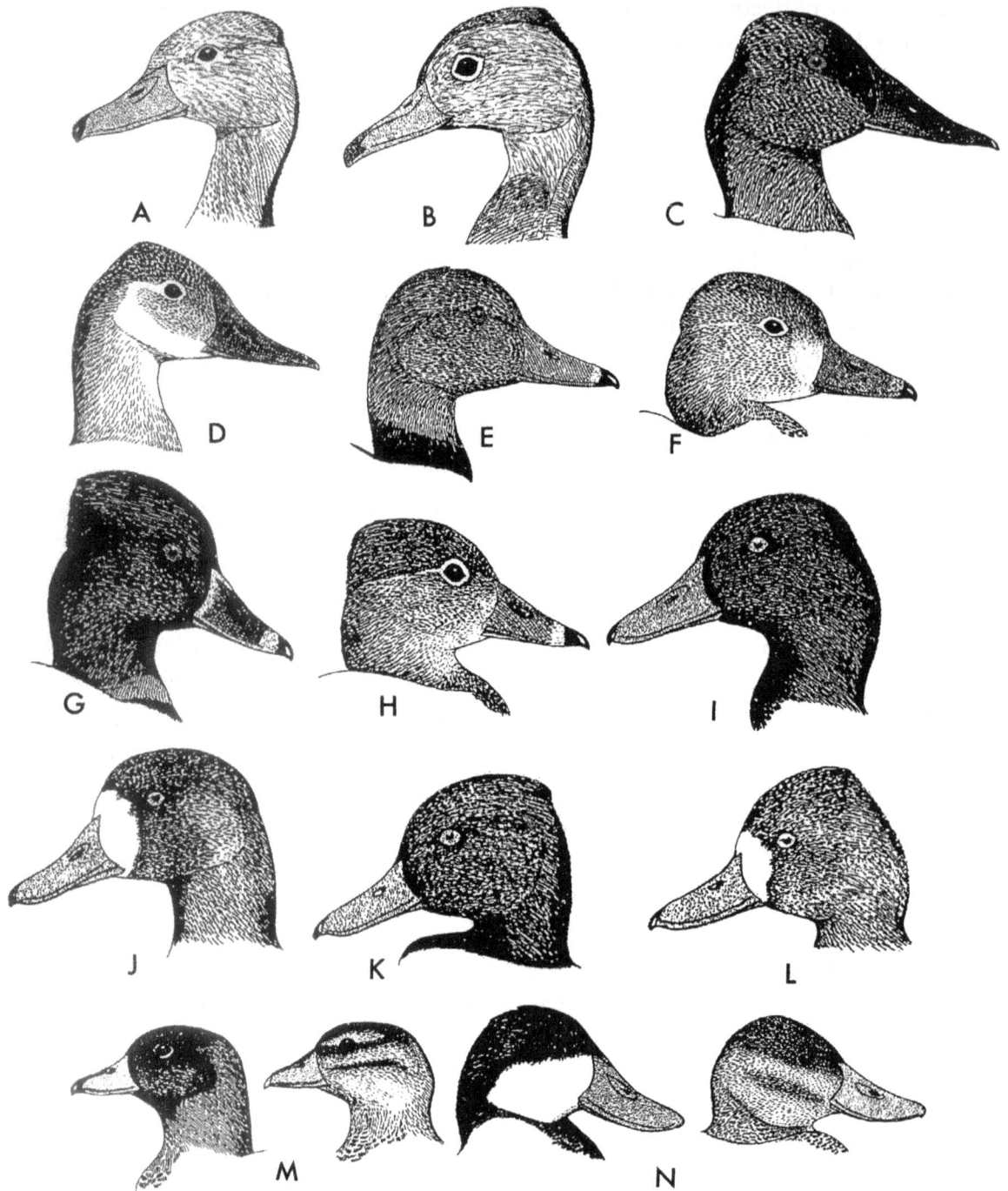

Fig. 2. Head profiles of breeding adult whistling-ducks, pochards, and stiff-tailed ducks, including (A) fulvous whistling-duck, (B) black-bellied whistling-duck, (C–D) male and female canvasbacks, (E–F) male and female redheads, (G–H) male and female ring-necked ducks, (I–J) male and female greater scaups, (K–L) male and female lesser scaups, (M) male and female masked ducks, and (N) male and female ruddy ducks.

Fig. 3. Maintenance and comfort behaviors of selected dabbling and diving ducks, including (A–C), preening, (D) yawning, (E) wing-and-leg stretch (F) two-wing stretch, (G) wing-flapping, (H) rotary head-shaking, (I) somersaulting, (J) swimming shake, (K) head-flicking, (L) bill-dipping, and (M) head-dipping.

and head-flicking is likewise used in the same context among stiff-tailed ducks. The universal general shake in display form is called the "upward shake" or "upward stretch." Other displays serve as threat or appeasement signals, as pair-forming or pair-bonding functions, and as family- and flock-integration mechanisms, such as providing pre-flight (flight-intention) or landing-coordination (settling) information.

Other general postures and functionally important movements that are used by a many waterfowl include sleeping (Fig. 4A), standing (Figs. 4D and 4F), dabbling at the water surface (Figs. 4B and 4C), tipping-up while foraging below the surface (Fig. 4E), and underwater locomotion (Fig. 4G). Regarding the latter, rather than using the alternate downward-oriented paddling strokes employed during surface swimming, the legs of stiff-tailed ducks are situated so far to the rear of the body that they can be raised to a position at right angles to the vertical body axis, allowing for a highly efficient simultaneous rowing by the two legs while submerged, as in loons and grebes (Johnsgard, 1987). Conversely, this adaptation reduces the stiff-tails' walking abilities on land, makes takeoffs from water more difficult, and makes land takeoffs difficult if not impossible.

Just as dabbling ducks use their wings to push them out of water during takeoffs, some diving ducks use a strong flick of their wings to help them submerge. Some also hold their wings slightly away from the body when swimming under water, perhaps for steering. However, no waterfowl are known to use their wings for underwater propulsion in the manner of auks and penguins.

In all these important biological attributes, the whistling-ducks, pochards, and stiff-tailed ducks offer three capsule versions of the remarkable morphological and behavioral diversity to be found among the entire family of waterfowl, and invite the reader to learn more about all of them.

Fig. 4. Resting and foraging behaviors of various dabbling and stiff-tailed ducks, including (A) sleeping (*Oxyura*),
(B) dabbling from land (*Stictonetta*), (C) dabbling while swimming (*Stictonetta*), (D) standing (*Heteronetta*),
(E) tipping-up (*Anas*), (F) standing (*Oxyura*), and (G) underwater swimming (*Oxyura*).

II. Species Accounts

Tribe Dendrocygnini (Whistling-Ducks)

The whistling-ducks comprise a distinctive worldwide group of nine species that are collectively of pantropical distribution, with only few species having ranges extending north to the Tropic of Cancer, and only two (the white-faced and fulvous whistling-ducks) occurring in both the Eastern and Western Hemispheres. In common with the swans and true geese (which with them compose the subfamily Anserinae), all the included species have a weblike (reticulated) tarsal surface pattern and lack sexual dimorphism in plumage. Their downy young have a unique plumage pattern that is generally ducklike but includes a pale band that extends back around the nape. Although their relationships are clearly with the geese and swans, whistling-ducks have sometimes been separated taxonomically as a distinct family Dendrocygnidae.

Whistling-ducks become sexually mature during their first year and are sexually monomorphic in plumage, with single annual molts. Like geese and swans, their social displays involve simple, mutually performed postures, often accompanied by vocalizations. Whistling-ducks form permanent pair-bonds and lack complex pair-forming displays. They also have a unique tracheal-syringeal structure and vocalizations that are very similar in both sexes. Their clear, often melodious, whistling voices of up to four notes are the basis for their group name, and all species are notably vociferous, especially in flight. Their alternative common name, tree-ducks, is far less appropriate since, although several species can perch on large branches, only a few consistently nest in tree cavities.

All whistling-ducks have relatively longer legs (the tarsal length is at least 20 percent of the wing length) and more rounded wings (the third, rather than the outermost, primary is the longest) than geese or swans, and they are substantially smaller, weighing up to about 1,200 g. All whistling-ducks call frequently while flying, and in flight their feet extend well beyond their relatively short tails, producing a unique flight profile. In spite of their long legs, some whistling-ducks dive surprisingly well and probably obtain most of their food in this manner. Except as ducklings, their foods are mostly or almost entirely from vegetative sources, and among several species much of their foraging is done nocturnally.

Whistling-ducks lack the triumph ceremonies that are so important in the pair-forming and pair-bonding behaviors of geese and swans. Also unlike geese, paired birds of at least four species perform mutual preening of the head and neck feathers (allopreening). Whistling-ducks are also unique in that the postcopulatory behavior of several species consists of a remarkable "step-dance," in which the pair rises in side-by-side unison with rapid foot-paddling and calling, as each bird lifts the somewhat folded wing opposite its partner to a vertical or near-vertical position.

Unlike geese and nearly all swans, both sexes of all species develop incubation patches and equally share both incubation and brood-rearing duties. In further contrast to both swans and geese, their nests typically lack any down insulation, but their tropical breeding distributions probably make this trait unnecessary. Incubation periods reportedly range from 22 to 31 days, and fledging periods are evidently relatively long—up to 65 days among the few species for which it has been reported.

Eight species are represented in the genus *Dendrocygna*, including all three of the species described in this book. A ninth species, the African and Madagascan white-backed duck (*Thalassornis leuconotus*), was historically considered to be a member of the stiff-tailed duck assemblage (Oxyurini), until I determined it to instead be an aberrant whistling-duck (Johnsgard, 1966). The highest degree of species diversity in *Dendrocygna* occurs in South America and Borneo, both of which support three species. No intertribal whistling-ducks hybrids have yet been documented, suggesting that the group is genetically well isolated from all other waterfowl groups.

Fulvous Whistling-Duck
Dendrocygna bicolor (Vieillot) 1816

Other vernacular names. Fulvous tree duck, long-legged duck, Mexican squealer

Range. Resident in Myanmar (Burma), India, Sri Lanka, Madagascar, and eastern Africa. Also resident in northern and eastern South America and from southern Mexico north to the southern United States, including Texas and Louisiana, where partly migratory, and Florida, where resident. Also resident in the Greater Antilles, where it is common in Cuba, less common in Hispaniola and Puerto Rico, and rare elsewhere.

Subspecies. None recognized by Delacour (1954). Few, if any, other birds in the world have such an enormous geographic distribution, ranging over four continents, without showing evidence of subspeciation.

Measurements. *Folded wing:* Kear (2005): Males 216–222 mm (ave. of 6, 219 mm); females 221–225 mm (ave. of 6, 223.3 mm). Hohman and Lee (2001): Males ave. at least 280, 212 mm; females, ave. of at least 280, 207 mm.

Culmen: Kear (2005): Males 47–49 mm (ave. of 6, 47.7 mm); females 45.3–48.9 mm (ave. of 6, 47.4 mm). Hohman and Lee (2001): Males, ave. of at least 280, 46.6 mm; females, ave. of at least 280, 45.5 mm.

Weights (mass). Hohman and Lee (2001): Males 545–958 g (ave. of 138, 771 g); females 595–964 g (ave. of 148, 743 g). Kear (2005): Males, ave. of 12, 637 g; females, ave. of 15, 614 g. Lynch (pers. comm.): Seven males (captives) 621–756 g, ave. 675.5 g (1.49 lb.); six females 632–739 g, ave. 689.9 g (1.52 lb.). Acosta Cruz et al. (1989): Males (Cuba), ave. of at least 280, 714 g; females, ave. of at least 229, 720 g.

Identification

In the hand. Like the other species in this genus, the presence of long legs that extend beyond the short tail, an entirely reticulated tarsus, and an elongated and elevated hind toe are distinctive. The fulvous whistling-duck is the only North American waterfowl with a grayish blue bill and foot coloration and extensive tawny-fulvous color on the head and underparts. The wings are entirely dark on the upper surface, without any white or pale gray patterning.

In the field. The most widespread species of whistling-duck in North America, fulvous whistling-ducks are likely to appear almost anywhere in the southern states. On water or land, their long and usually erect necks, ducklike heads, and short-tailed appearance are distinctive. At any distance, the fulvous whistling-duck appears mostly tawny brown, darker above and brighter below, with the buffy yellow flank stripe the most

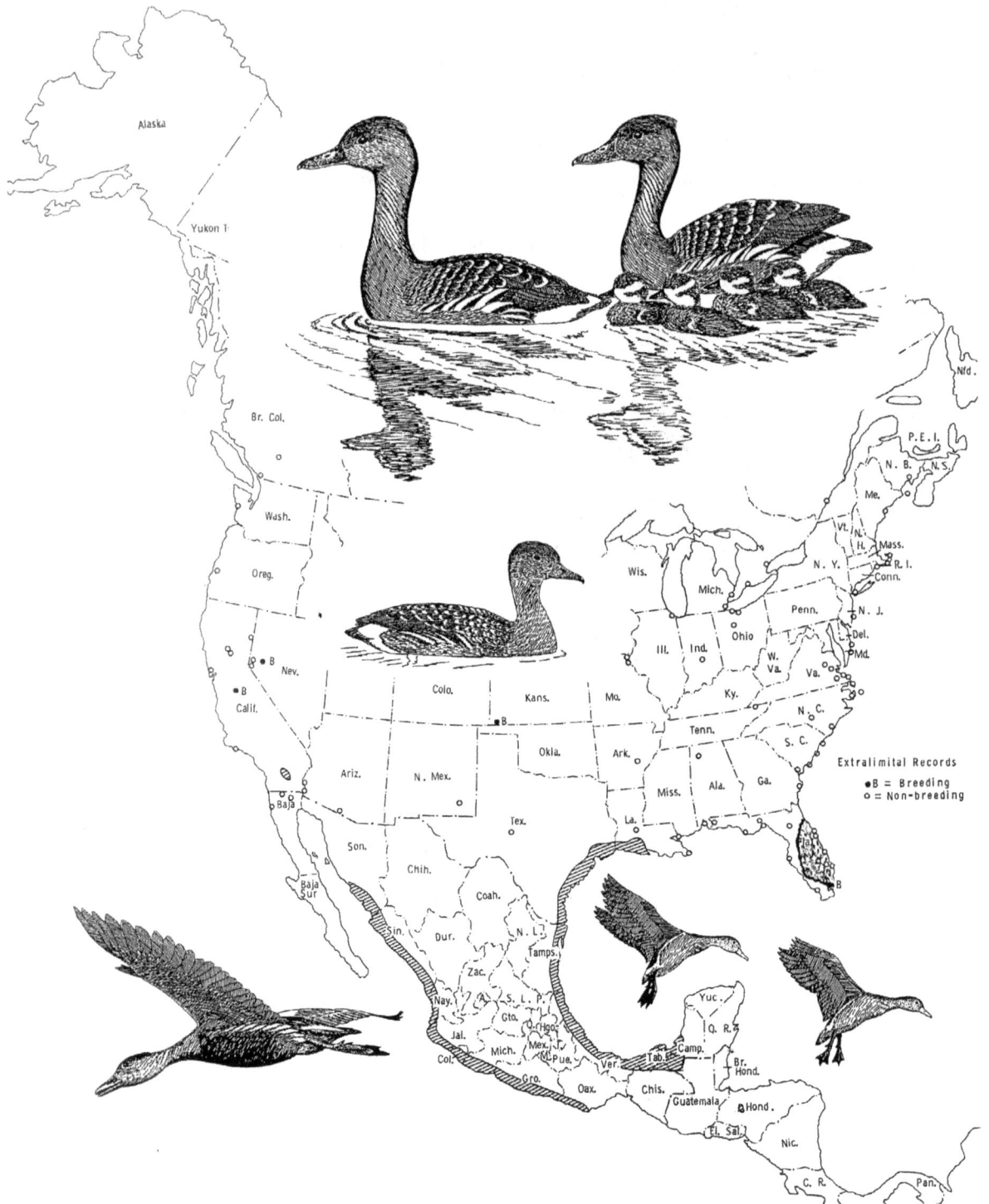

The breeding (hatched, with denser concentrations inked), wintering (shaded), and acquired (stippled) range of the fulvous whistling-duck.

conspicuous field mark. In flight, the long neck and long, often-dangling legs are evident, and the head is usually held at or even below the body level. In contrast to the wing coloration of the other two species of whistling-ducks that might be encountered in North America, the upper-wing surface is neither white nor grayish white but is instead dark brown like the mantle. The wings are broader and more rounded than in more typical ducks, and a distinctively slower wingbeat is characteristic. A whistled *wa-chew'* or *pa-cheea* call is frequently uttered, both in flight and at rest. The fulvous whistling-duck feeds in rice fields and shallow marshes and occasionally comes into cornfields as well.

Age and Sex Criteria

Sex determination. No obvious external sexual differences occur, so internal examination might be required. McCartney (1963) believed that females could be distinguished on the basis of being smaller, duller, and having a continuous rather than an interrupted dark line on the crown and neck. Acoustic analysis has shown that the sexes of adults can be distinguished by their voices, the male being lower in pitch than the female (Voliodin et al., 2009).

Age determination. Age determination is not yet well studied, but if the findings of Cain (1970) on the black-bellied whistling-duck apply, notched tail feathers might persist until about the 35th week of age, and the penis of a male under ten months of age lacks spines. Dickey and van Rossem (1923) suggested that immature birds could be distinguished from older ones by the former's concave rather than straight bill profile; they also indicate the plumage of immature birds is very similar to that of adults, but the brown tips on the back feathers average slightly darker.

Distribution and Habitat

Breeding distribution and habitat. During the early part of the twentieth century, the fulvous whistling-duck was believed to be limited as a breeding species to Texas and central and southern California, with possible casual breeding in central Nevada, southern Arizona, and Louisiana as well (Bent, 1925). In the mid-1960s the first Florida breeding record was obtained at Lake Okeechobee, where the population soon grew to about 200 birds. Following the development of large winter flocks in the vicinity of Virginia Key, Dade County, breeding was verified there in 1968, and nests or broods were consistently found thereafter. By 2016 the species had expanded its residential range to include essentially all of peninsular Florida.

Breeding in Texas began as early as about 1916, and by the 1970s the species was breeding along the south Texas coast in the vicinity of San Beni, Brownsville, and the Santa Ana National Wildlife Refuge, north locally through the Corpus Christi area, and inland as far as Mathis. By 1930 the species was abundant in the east Texas rice belt as far west as Colorado County, according to Carroll (1932), who first related the bird's distribution and abundance in Texas to the developing rice-growing culture. By 2010 the species was a common summer resident in the coastal prairies and the eastern half of the lower Rio Grande valley north and east to the Louisiana border, and was a rare to locally uncommon winter resident north to Baffin Bay (Lockwood and Freeman, 2014).

Nesting in Louisiana was first verified in 1939 (Lynch, 1943), and there it was later determined to be a common breeding bird in the rice belt (Meanley and Meanley, 1959). By 2000 the species was breeding commonly in rice-growing areas east to the St. Landry, St. Martin, and Iberia parishes (Hohman and Leer, 2001). Most of the Louisiana population is migratory, but some birds overwinter.

Although many California localities had breeding records as far north as San Francisco Bay during the early 1900s (Grinnell and Miller, 1944), breeding in California was highly localized by the 1970s, as a probable result of changing agricultural practices and overhunting, and breeding was increasingly confined to the vicinity of Los Banios, Merced County. Later the birds survived in very small numbers in the Salton Sea region, where as of 2001 a few pairs still persisted at the southern end of the Salton Sea (Hohman and Lee, 2001; Hamilton, 2008). The historic breeding habitat in California consisted of freshwater marshes where tules or cattails grow interruptedly (Grinnell and Miller, 1944).

An interesting recent phenomenon has been the proliferation of postbreeding (fall and winter) records of fulvous whistling-ducks in the eastern United States and, to a more limited extent, in the central and western states as well. Hartz (1962), Jones (1966), and Bole and Rylander (1983) summarized the early records. Jones plotted on a map the records he found for the period 1949–65; these have been transferred to the accompanying range map, and some additional or more recent records have also been added. Since then far too many extralimital records have accrued to map them; for example, fulvous whistling-ducks were reported from Kansas during 14 different years from 1929 to 2011 (Thompson et al., 2011), and there were sight records from nearly all states and provinces by 2014 (Baldassarre, 2014).

Population. Neither the annual Audubon Christmas Bird Counts nor the US Fish and Wildlife Service's Midwinter Waterfowl Surveys provide much information on the total US population of fulvous whistling-ducks. The total US Audubon Christmas Counts averaged fewer than 800 birds annually for the period 2005–06 to 2009–10, probably reflecting a general migration out of Louisiana and Texas to Mexico. However, Hohman and Lee (2001) reported that at Loxahatchee National Wildlife Refuge, Florida, peak counts increased from less than 300 birds in the 1960s to nearly 11,000 in 1990. No current statewide population estimates are available for Florida, Louisiana, or Texas, but Anderson, Muehl, and Tacha (1998) estimated Texas totals of 18,000 to 22,500 birds, observed during fall and winter surveys in 1992–93.

Hunter kills are still small; from 1961 to 1998 they annually ranged from 200 to 4,600 birds for the entire United States (Hohman and Lee, 2001). From 1998 through 2007, US hunters killed an estimated total of 11,000 birds (Baldassarre, 2014). The species is not individually identified in the most recent (2013–15) annual report on migratory bird hunting in the United States (Raftovich, Chandler, and Wilkins, 2015).

Wintering distribution and habitat. Considerable seasonal movements are typical of this species, and it is thought that the majority of the Louisiana-Texas population moves to Mexico during the winter. Leopold (1959) remarked that in Mexico the largest winter populations occur in coastal Guerrero, although

Fig. 5. Fulvous whistling-duck, adult swimming

the species is not abundant even there. There is also an apparently sedentary Mexican population that occurs on the coasts of Sonora, Sinaloa, Nayarit, and Guerrero and on the Caribbean coasts of Tabasco, Veracruz, and Tamaulipas, which are probably enhanced to some degree by winter migrants.

Howell and Webb (1995) noted that the species' Mexican distribution is nomadic and unpredictable. Midwinter Waterfowl Surveys taken on the Gulf coast of Mexico since 2000 found the great majority of birds (up to a few thousand) on the Tabasco lagoons, with relatively few having been seen in the interior highlands, and even fewer along the Pacific coast (Baldassarre, 2014).

General Biology

Age at maturity. The usual age of sexual maturity is still somewhat uncertain, but inasmuch as captive birds sometimes breed during their first year, it might be assumed that this at least occasionally occurs in the wild. Marvin Cecil stated (pers. comm.) that to his knowledge the fulvous whistling-duck is the only species of whistling-duck that often breeds in its first year of life, whereas the others do not breed in captivity until their second year. Meanley and Meanley (1958) observed normal copulation by a male when it was eight months old. McCartney (1963) suggested that yearlings might be relatively late nesters, judging from his observations of captive birds.

Pair-bond pattern. Whistling-ducks have strong pair-bonds, with the male regularly assisting in the rearing of the young. For this reason it is assumed that the normal pair-bond is permanent, as in geese and swans, although actual data on this point appear to be lacking. Although pairing might normally be permanent, some pair-bonds are formed after the birds arrive in spring (Meanley and Meanley, 1958), which might reflect pairing by first-time breeders, or perhaps result from the need for re-pairing by individuals that had lost their prior mates.

Nest location. Dickey and van Rossem (1923) reported that all of "some 50" California nests they located in 1921 were located in tufts of a dwarf species of *Scirpus*, although in 1922 these tules were flooded and nests occurred in dense clumps of living or dead *Scirpus* of a larger species, in knotweed (*Polygonum*), or on floating materials in open water. Lynch (1943) commented that Louisiana nests were usually found in rice fields, on levees or along dikes, or sometimes built as floating structures in standing rice. Meanley and Meanley (1959) noted many nests were located on rice field levees or built over water between levees, but others were attached to growing plants. At the Welder Wildlife Foundation in Texas the nests of this species were placed over water usually 3 to 7 feet in depth, among dense stands of grasses and cattails (Cottam and Glazener, 1959). In Louisiana rice fields, the densities of nests in one study averaged 15.1 per square kilometer (Hohman et al., 1994).

Clutch size. Because of the prevalence of "dump-nesting" by other females, the typical clutch size is difficult to ascertain. Dickey and van Rossem (1923) estimated the normal range to be 10 to 16 eggs, Lynch (1943) estimated 10 to 15, and Meanley and Meanley (1959) judged that 13 eggs represent an average clutch size. The average clutch size of 9 successful nests observed by Cottam and Glazener (1959) was 12.6 eggs. Hohman and Lee (2001) reported an average of 14.1 eggs for 193 ground nests, and 13.4 for 296 nests situated over water. The rate of egg laying is apparently one per day (Meanley and Meanley, 1959; Dickey and van Rossem, 1923).

   Intraspecific brood parasitism, or "dump-nesting," is common in this species; brood parasitism rates of 34 percent among 193 upland nests, and 42 percent of 489 rice-field nests have been reported (Hohman and Lee, 2001). The eggs of at least three other duck species (northern pintail, redhead, and ruddy duck) have also been found in fulvous whistling-duck nests.

Incubation period. The incubation period is normally 24 to 26 days, with estimates of 24 days by Meanley and Meanley (1959), 25 days by Dickey and van Rossem (1923), and 28 days by Johnstone (1970). Delacour's (1954) unusually long estimate of 30 to 32 days might have resulted from eggs laid by multiple females; one clutch of 22 eggs reportedly had an incubation period of 42 days (Gómez Ventura and de Mendoza, 1982).

Fulvous whistling-duck, pair and ducklings

Fledging period. Meanley and Meanley (1959) noted that initial flight occurred in a captive female at 63 days. No other estimates are available.

Nest and egg losses. A high incidence of nest losses by desertion or by flooding was noted by Dickey and van Rossem (1923), and likewise Meanley and Meanley (1959) suggested that initial nesting success was apparently low, with only 3 of 10 observed nests being successfully hatched. Cottam and Glazener (1959) reported that 9 of 17 nests they studied were successful, and 94 of 164 eggs were hatched, a hatching success rate of 57 percent. In the 9 successful nests, 94 of 113 eggs hatched, an 83.2 percent hatching success rate.

Renesting probably compensates for breeding failures and is facilitated by a prolonged breeding season (Hohman and Lee, 2001). Nests have been found as late as August in both Louisiana and California, and in Texas there are egg records from May 16 to September 19 (Bent, 1925), indicating a breeding season of nearly four months. In spite of this long breeding period the birds are single-brooded, with renesting likely in the case of failed clutches or possibly lost broods.

Juvenile and adult mortality. There are no available estimates of mortality rates in this species, although many writers have commented on its susceptibility to hunters because of the species' unwary behavior and their fragile bone structure. Meanley and Meanley commented that since the ducks are so readily killed, it is fortunate that most of them have moved southward out of Louisiana prior to the start of the waterfowl hunting season.

General Ecology

Food and foraging. Few studies on the foods and feeding behavior of fulvous whistling-ducks have been performed, but they all indicate a high dependence on plant matter. Howard Leach (cited by Leopold, 1959) found that in the crops of five birds taken in California's Imperial Valley, the seeds of water grass (*Echinochloa*) predominated, with small quantities of *Polygonum* and *Melilotus* also present. From stomach analysis Dickey and van Rossem (1923) noted that wild timothy (*Phleum*) formed the bulk of the summer food during one year, whereas the seeds of *Polygonum* species were important in the late summer and fall of 1922.

Meanley and Meanley (1959) determined that rice seeds composed 78 percent of the food of 15 birds collected in water-planted rice fields near the coast, whereas in dry-planted fields and in early fall samples rice was a minor part of the diet, with weed seeds forming the bulk of the food.

When foraging, the birds often pull down the seed heads of emergent plants and strip them. They also often feed by tipping-up, or simply by lowering the head into the water without tipping-up. They also dive well and may remain submerged from about 9 to 15 seconds, with intervening surface periods of 10 to 18 seconds (Johnsgard, 1967b). Although ill-adapted for perching, this species is better adapted to diving than are black-bellied whistling ducks, and fulvous whistling-ducks have been seen foraging in water as deep as 170 centimeters (67 inches) (Baldassarre, 2014).

Meanley and Meanley (1959) performed studies on possible depredations on rice crops but found little evidence of significant damage to rice by this species, a conclusion that was also reached by Hohman et al. (1996).

Sociality, densities, territoriality. The extreme sociality of this species has been stressed by Dickey and van Rossem (1923), who mentioned that even during the peak of the laying season the birds continually gathered into small groups of mated pairs for feeding and resting together, separating only in the early morning hours for laying. Several larger flocks, apparently of nonbreeding birds, were also present through the summer period, reaching a minimum in early July and then being augmented by apparently unsuccessful nesters.

Such sociality sometimes favors high nest concentrations, at least when favored nesting habitat is restricted. Dickey and van Rossem noted about 50 nests in an area approximately half a mile long by 200 yards wide and felt that many more nests were present but remained undetected. These figures would suggest a nesting density of at least 1.4 nests per acre; the equivalent of nearly 900 nests per square mile. Nest concentrations (averaging 15.1 per square kilometer, or 54 per square mile) have been found in rice fields

Fulvous whistling-duck, adult resting

(see the "Nest location" section). Meanley and Meanley (1959) found a much lower breeding density of 1.3 to 2 pairs per square mile in two 5-square-mile study areas.

Interspecific relationships. It is possible that some competition for food exists between the fulvous and black-bellied whistling-ducks, but since their nest site preferences are wholly different there would seem to be little if any competition for breeding locations. Rylander and Bolen (1970) also pointed out that whereas the black-bellied whistling-duck is primarily a wading and perching species, the fulvous is mainly a swimming forager and mostly dabbles for food. Rylander and Bolen related the relatively larger foot size of the fulvous to the fact that it is a better swimmer and has probably greater reliance on diving while foraging.

Nesting associates of fulvous whistling-ducks in Louisiana include such typical coastal marsh birds as the red-winged blackbird, purple gallinule, king rail, least bittern, and marsh wren (Meanley and Meanley, 1959). As noted earlier (see the "Clutch size" section), northern pintail, redhead, and ruddy duck eggs

Fig. 6. Behavior of fulvous whistling-duck (A–F), black-bellied and West Indian whistling-ducks (G), and West Indian whistling-duck (H), including (A) landing, (B) precopulatory mutual head-dipping, (C–E) postcopulatory step-dance, (F) head-low-and-forward threat, (G) swimming threat, and (H) postcopulatory display (after Johnsgard, 1965).

have been found in California nests containing those of fulvous whistling-ducks (Dickey and van Rossem, 1923), and all three of these species are known to be brood parasites elsewhere (Weller, 1959).

General activity patterns. The nocturnal foraging activities of whistling-ducks are well documented, although daytime foraging is also common. Meanley and Meanley (1959) noted that in late April fulvous whistling-ducks usually would leave the coastal marshes about 8:00 p.m. for the rice fields, often in flocks of 30 to 40 birds. Later in the summer, flocks of 150 to 200 birds were seen by them in rice fields, and a maximum of 3,000 birds were seen at Lacassine National Wildlife Refuge, Louisiana, in late summer. Cottam and Glazener (1959) suggested that this species' migration might also occur at night, which seems very likely.

Social and Sexual Behavior

Flocking behavior. The strong flocking behavior of this species, even during the breeding season, has already been noted. Because of their vociferous and strongly gregarious tendencies, fulvous whistling-ducks decoy readily and can be attracted to a whistled imitation of their call.

Pair-forming behavior. Presumably because of the strong and apparently persistent pair-bonds of this species, descriptions of pair formation are almost nonexistent. Meanley and Meanley (1959) noted what appeared to be courting flights in spring, when three or four ducks flew in unison during erratic flights. On one occasion a single female was observed being followed by three males on the ground. Observations on captive birds suggest that the male pair-forming displays are almost identical to those of geese, although triumph ceremonies are lacking (Johnsgard, 1965).

Copulatory behavior. Meanley and Meanley (1958) and Johnsgard (1965) described the copulatory behavior of this whistling-duck. This species typically copulates in water of swimming depth, and precopulatory activities are scarcely separable from normal bathing movements, involving head-dipping on the part of both birds (Fig. 6B). The postcopulatory "step-dance" (Fig. 6C–D) is a highly ritualized display in which both birds rise in a parallel fashion in the water, and each bird raises the folded wing on the opposite side from its partner as they both tread water rapidly.

Nesting behavior. Although nest locations vary considerably according to local conditions, they are typically in emergent vegetation and often are roofed over and nearly hidden from above. Nests in water often have ramps, sometimes several feet long, leading to the rim, and rarely if ever is any significant amount of down present in the nest. Males presumably help females construct the nest, and Delacour (1954) believed that the male might spend more time than the female at the nest.

Brooding behavior. Both sexes attend the young. They probably undergo their postnuptial molt and flightless period of a few weeks at about the same time, during the two-month-long fledging period. McCartney (1963) noted that most hatching dates in Louisiana occurred in July, but the adults' peak flightless period was in mid-September.

Postbreeding behavior. With the fledging of the young, families gather into larger flock units and, after molting, move to favorable feeding areas prior to the fall migration. Dickey and van Rossem (1923) noted that, although in 1921 all the birds had left Buena Vista Lake by the first of September, in 1922 favorable water conditions attracted "thousands" of birds, which began to move south shortly after the first of October. McCartney (1963) suggested that the eastern Texas and Louisiana population migrates nonstop to and from Mexican wintering grounds on Mexico's Gulf coast, an air distance of about 600 miles. A similar migration has since been confirmed for Florida birds as well as for over-water flights between Florida and Cuba (Flickinger, King, and Heyland, 1973).

West Indian Whistling-Duck
Dendrocygna arborea (Linnaeus) 1758

Other vernacular names. Antillean tree duck, black-billed tree duck, Cuban whistling-duck

Range. Resident on some of the Bahamas (Andros, Exuma, Great Inagua) and the Greater and Lesser Antilles, including Antigua, Barbados, Cuba, Grand Cayman, Hispaniola, Isla des Juvenud, Jamaica, Puerto Rico, and (formerly) the Virgin Islands (American Ornithologists' Union, 1998).

Subspecies. None recognized.

Measurements. *Folded wing:* Kear (2005): Males 250–284 mm (ave. of 18, 268 mm); females 243–274 mm (ave. of 11, 264.7 mm).
   Culmen: Kear (2005): Males 50–55.5 mm (ave. of 18, 53.4 mm); females 47.7–54.3 mm (ave. of 11, 52.0 mm).

Weights (mass). Kear (2005): Males 760–1,240 g (ave. of 18, 984 g); females 860–1,320 g (ave. of 11, 1,064 g).

Identification

In the hand. Identifiable as a whistling-duck on the basis of the long legs, entirely reticulate tarsus, and the elongated hind toe, this species is the largest of all whistling-ducks. Its folded wing measurements (230–270 mm) and its long, black bill (culmen 45–53 mm) separate it from all other species of the genus.

In the field. This West Indian species is unlikely to be seen in continental North America, except as possible escapes from captivity. It is the only Northern Hemisphere whistling-duck that is predominantly dark brown, with a blackish bill and mottled black and white flanks. Like the other whistling-ducks, it has long legs, a short tail, and rounded wings, which produce a distinctive, somewhat gooselike, body profile. Its call is rather infrequently uttered but is a clear whistle sounding like *wheet-a-whew'-whe-whew'*. The birds swim well, often foraging in shallow waters or on dry land. This species also perches in trees to some extent. Nesting is done in tree cavities, on tree branches covered with vegetation such as bromeliads, or on the ground. In flight, the species exhibits ashy-white markings on the wings in the areas where the black-bellied whistling-duck has pure white feathers.

Fig. 7. West Indian whistling-duck, adult standing

Age and Sex Criteria

Sex determination. No plumage characteristics are available to distinguish the sexes externally. Acoustic analysis has shown that the sexes of adults can be distinguished by their voices, the male being higher in pitch than the female (Voliodin et al., 2009) when the two birds are duetting (Fig. 6H).

Age determination. Juveniles birds are duller and grayer than adults, with gray rather than white upper tail-coverts. Tail molt has not yet been analyzed, but the distinctive notched juvenal tail feathers are no doubt carried for much of the bird's first fall of life.

Occurrence in North America

According to Bond (1971), this species' major range includes the Bahamas, the Greater Antilles, and the northern Lesser Antilles, but it is of only casual occurrence elsewhere in the West Indies. There have been several recent (2003, 2007, 2014) eBird reports from Andros Island, Bahama Islands. The species has not yet been adequately documented from continental North America, but there was an occurrence of a bird at the Great Dismal Swamp, Virginia, that was photographically documented (in eBird) during April 2003. There have also been sightings reported from Florida that might have represented vagrants from Cuba. The species is now rare over most of its range and has been classified as Vulnerable by BirdLife International.

Black-bellied Whistling-Duck
Dendrocygna autumnalis (Linnaeus) 1758

Other vernacular names. Black-bellied tree duck, gray-breasted tree duck, pichichi, red-billed tree-duck, red-billed whistling-duck

Range. From northern Argentina northward through eastern and northern South America, Central America, Mexico, and the southern United States along the Gulf coastal plain from the Texas-Mexican border to southwestern Louisiana, throughout peninsular Florida, and southern Arizona.

North American subspecies. *D. a. autumnalis* (L.): Northern black-bellied whistling-duck. Ranges from North and Central America south to Panama, where it is replaced southwardly by *D. a discolor*, the southern black-bellied whistling-duck.

Measurements. *D. a. autumnalis: Folded wing:* Bolen (1964): Both sexes 229–249 mm (ave. of 21, 268 mm).
Culmen: Bolen (1964): Both sexes 39–56 mm (ave. of 21, 53 mm).

Weights (mass). Bolen (1964): Males 680–907 g (ave. of 35 in May, 984 g); females 652–1,021 g (ave. of 39 in May, 839 g). Nine breeding-season males 728–952 g, ave. 799.5 g; eight breeding-season females 832–978 g (ave. 893.4 g).

Identification

In the hand. Like the other whistling-ducks, this species has long legs that extend beyond the short tail in flight, an entirely reticulated tarsus, and an elongated and elevated hind toe. It is the only whistling-duck with a red bill, pink feet, and white on the upper wing-coverts.

In the field. Whistling-ducks stand in a rather erect posture on land, where their long necks, long legs, and duck-like body are evident. In the water they swim lightly, with the tail well out of the water and the neck usually well extended. The black-bellied whistling-duck is easily recognized in both situations by its red bill and the large white lateral stripe of wing feathers that separates the brownish back from the black underparts.

In flight, the long neck and trailing legs are apparent, and the blackish underparts and underwing surface contrast strongly with the predominantly white upper-wing surface. Both in flight and at rest, the birds often utter a series of clear whistling notes, the most typical of which is a 4- to 7-note call sounding like *wha-chew'-whe-whe-whew* or *pe-cheche-ne* (Leopold, 1959). As a cavity-nesting species, it is often seen perching in trees. It is highly gregarious, and gathers in large flocks when not breeding.

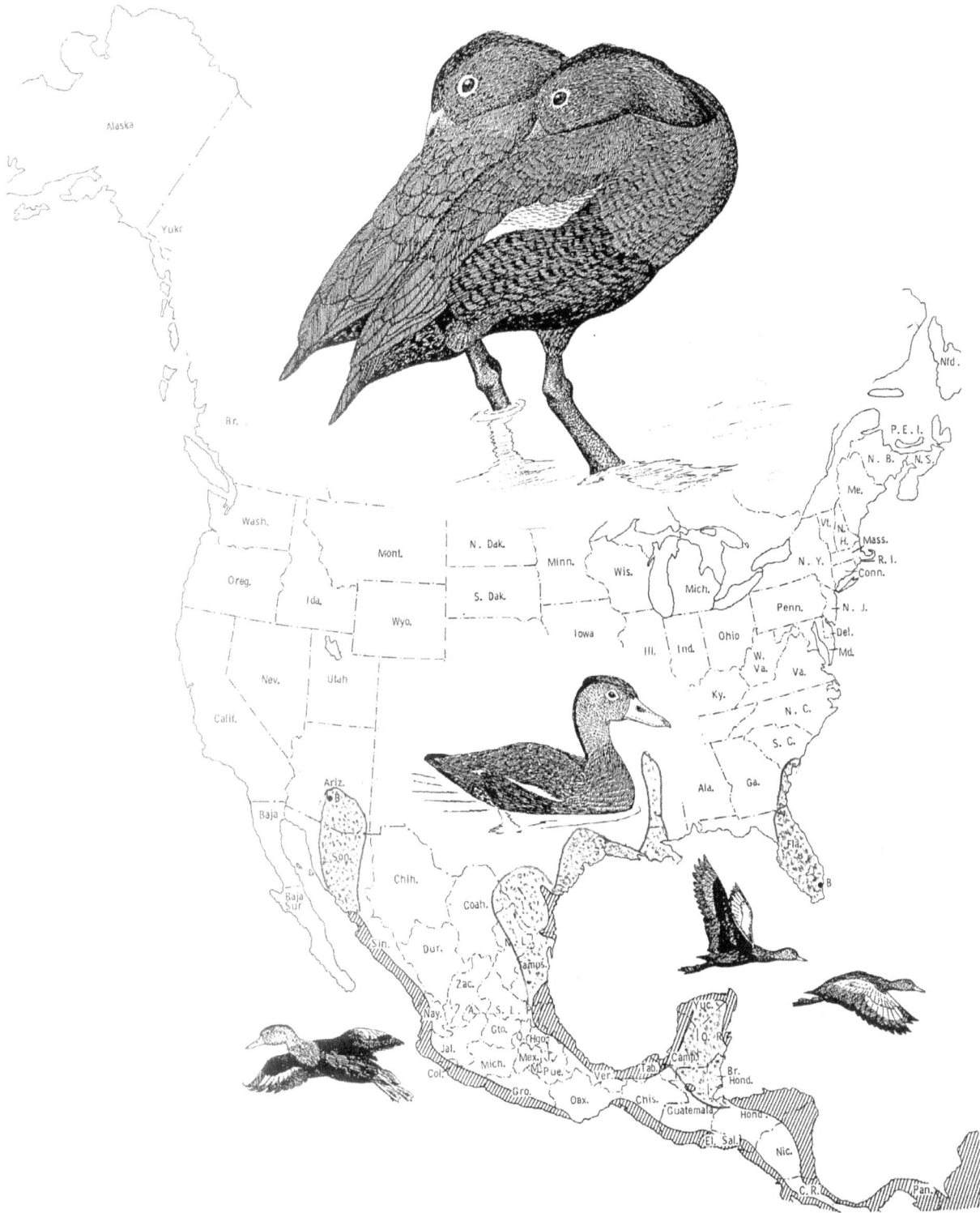

The breeding (hatched, with denser concentrations inked), wintering (shaded), and acquired (stippled) range of the black-bellied whistling-duck.

Age and Sex Criteria

Sex determination. There are no apparent external differences in the sexes, so internal examination is required for determination of sex. Acoustic analysis has shown that the sexes of adults can be distinguished by their voices, the male call being longer in duration than the female (Voliodin et al. 2009).

Age determination. Juveniles are generally more grayish than adults, with gray heads, grayish legs and feet, and a pale pink to grayish bill color. According to Cain (1970), notched juvenal rectrices might persist until the bird is about 9 months old. On birds 6 to 8 months old, the black feathers of the rump region are tipped with white, and the penis of males lacks spines, whereas those at least 10 months old have entirely black rump feathers, and males have well-developed spines on the penis.

Distribution and Habitat

Breeding distribution and habitat. In the United States, the historic breeding area of the black-bellied whistling-duck was historically almost entirely limited to southern Texas. In 1962 Bolen stated that the northernmost part of its breeding range was within a 50-mile radius of Corpus Christi. It had by then been reported as far north as Eagle Lake (Peterson, 1960), and Bolen et al. (1964) then considered it to be "well established" in Live Oak, San Patricio, Kleberg, and Brooks Counties. By then it was also a common breeder in the lower Rio Grande valley, including Santa Ana and Laguna Atascosa refuges. As of 1975 it had bred in the vicinities of Rio Hondo, Brownsville, and Falfurrias, and broods or nests had been found north of Corpus Christi at Mathis, Beeville, and the vicinity of San Antonio. In some years as many as 20 pairs were then nesting at Santa Ana Refuge, and several hundred young had been seen in favorable years at Laguna Atascosa National Wildlife Refuge.

As a result of irrigation development and stock-pond construction, the birds rapidly moved north and east, eventually reaching southwestern Louisiana, in part as a result of releases from wildlife refuges (Wiedenfeld and Swan, 2000). Inland breeding in Texas has continued to increase, and nestings have been reported north at least as far as Dallas (Reinking, 2004; Baldassarre, 2014). Birds breeding in southern Texas are year-round residents, but more northern populations are migratory and usually leave Texas by December.

By 2000 the species had colonized the southwestern corner of Arkansas and north along the Mississippi River valley region as far as Tennessee. Breeding has been documented in Arkansas, the birds very possibly having originated in Texas (James and Thompson, 2001). Initial Tennessee breeding was reported in 1998 (Baldassarre, 2014), and breeding in Oklahoma was first reported in 1999 (Kamp and Loyd, 2001; Reinking, 2004). Post-breeding wandering is common, with some birds reported from as far north as Vancouver Island, Canada.

Meanley and Meanley (1958) described the nesting environment of this species in Texas. They found ten nests in a thicket of trees and shrubs near a small lake. All the nests were in hollow trees, eight of which were ebony (*Pithecolobium*) and two hackberry (*Celtis*).

Fig. 8. Black-bellied whistling-duck, adult wading

Outside of Texas, only a few breeding records had been reported for the United States before 1970. There were two breeding records for the Miami area, which might have represented escapes from the Crandon Park Zoo. However, most Florida breeding birds might have originated from wild birds from Mexico that had established a breeding center near Sarasota (James and Thompson, 2001). This population expanded to eventually include all of Florida except for the panhandle region (Bergstrom, 1999). The population had been seen farther north into South Carolina by 1994 (Harrigal, Laurie, and Floyd, 1995) and had begun nesting in that state by 2003 (Harrigal and Cely, 2004). Color banding of South Carolina birds in 2014 resulted in sightings from Florida, Georgia, Virginia, and New Jersey within a year.

The first definite record of nesting in Arizona was obtained near Phoenix in 1969 (Johnson and Barlow, 1971), although for several years the species had been increasingly seen around Phoenix, Tucson, and Nogales. After becoming established in the Santa Cruz valley, the birds expanded into central and southeastern Arizona (Brown, 1985).

In Mexico this species is much more common than the fulvous whistling-duck, breeding principally along the tropical coasts but occasionally nesting in the temperate uplands as high as about 2,400 feet (Leopold, 1959; Howell and Webb, 1995). The northern race of the black-bellied whistling-duck also breeds commonly farther south in Central America to central Panama and is replaced in South America by another race (*D. a. discolor*) with a more grayish breast color.

Population. US Fish and Wildlife Service Midwinter Waterfowl Survey data reported an average of 58,000 black-bellied whistling-ducks along the Gulf coast of Mexico between 1978 and 1994, compared with 14,000 on the Pacific coast, and less than 100 in the interior highlands. More than 85 percent of the birds on the Gulf coast were seen in Veracruz between the Tamiahua Lagoon (south of Tampico) and the city of Veracruz (Baldassarre, 2014). Probably most of these migrants were from Texas and Louisiana. The same winter survey route reported 81,000 birds along the coast of Texas in 2008.

Birds overwintering in Florida have been observed on the annual Audubon Christmas Bird Counts since the early1970s. During the 2009–10 counts, a total of more than 20,000 black-bellied whistling-ducks were tallied, of which 73 percent were seen in Texas, 20 percent in Florida, and 6 percent in Louisiana. In the 2010–11 hunting season, the total US kill of 17,500 birds were nearly all shot in Texas (40 percent), Florida (37 percent), and Louisiana (23 percent) (Baldassarre, 2014).

Wintering distribution and habitat. In southern Texas this species was usually present from April to early November until the early 1960s, with only a few birds normally overwintering (Bolen, 1962). More recently some of the Rio Grande delta birds have become residents (Reinking, 2004). It appears that the northern Texas and Louisiana populations move into the coastal regions of eastern Mexico. Leopold (1959) mentioned large winter flocks in the mangrove swamps of Nayarit, and smaller numbers of both species of whistling-ducks were noted in the rivers and lagoons of Veracruz and Tabasco. Reportedly this species also at times occurs in large numbers on the south coast of Chiapas as well as on the larger rivers in the northern part of that state.

There have been only a few band recoveries that provide information on migration routes and distances. Bolen (1967a) reported that the bands from eight of nine birds banded near Corpus Christi were recovered in Tamaulipas, Mexico, but the other band was recovered in San Luis Potosi. Out of a sample of 1,807 birds banded in Texas, 25 were out-of-state recoveries, of which 20 were from Louisiana and 5 from Mexico (Baldassarre, 2014).

Black-bellied whistling-duck, pair swimming

General Biology

Age at maturity. Not established with certainty, but males develop spines on the penis and acquire a fully adult plumage by 10 to 21 months of age (Cain, 1970), suggesting that breeding initially occurs at the end of the first or second year of life. Ferguson (1966) stated that two of six aviculturists responding to a survey reported initial breeding in each of the first three years of life.

Pair-bond pattern. Like the other species of *Dendrocygna*, this species exhibits a strong pair-bond, with the male assisting in nest and brood defense. There is definite evidence (Bolen, 1967b) that the male participates in incubation. The pair-bond is evidently permanent and potentially lifelong (Bolen, 1971). There are several known cases of pairs remaining intact up to four years. However, there have also been a few cases of broken pair-bonds for which both pair members were still alive; pair-bonds were later formed by them with new partners (Delnicki, 1983).

Black-bellied whistling-duck, pair and ducklings

Nest location. In contrast to the fulvous whistling-duck, this species preferentially nests in tree cavities or other elevated cavities, such as nest boxes. No down is normally present. Of 20 natural nest sites studied by Bolen et al. (1964), 17 were in trees and 3 were on the ground. Ten of the tree sites were water-isolated, and 5 were within 50 feet of water, but 2 were about a quarter mile from water. The occurrence of herbaceous rather than shrubby vegetation under the nest entrance might be important in nest site selection, as is the presence of a nearby perch. The height of the nest entrance averaged 270 centimeters (106 inches) for those tree nests situated above water, and about 160 centimeters (62 inches) for those over land. The ground nests consisted of shallow baskets made of woven grasses.

Clutch size. Eggs are laid one per day (Bolen, 1962). Bolen (1962) estimated the average clutch to range from 12 to 16 eggs. However, clutch size data obtained from casual field observations are obscured by a strong tendency for dump-nesting (brood parasitism) by this species. Bolen et al. (1964) observed that

Black-bellied whistling-duck, pair and ducklings

nearly half of 428 eggs found in southern Texas remained unhatched, apparently because of desertion related to multiple-bird nest use. McCamant and Bolen (1979) found that 70 percent of 778 nests studied had at least 14 eggs present, a total highly suggestive of brood parasitism. In later studies, by using egg measurement characteristics, James (2000) determined that among 134 nests examined all of the nests had evidently been affected by parasitism, with only 6 to 7 eggs having typically been laid by the host female.

Incubation period. Bolen et al. (1964) determined the average incubation period as 28 days. Cain (1970) stated that in an artificial incubator the eggs usually hatched 29 to 31 days after incubation began. Incubation periods ranged from 25 to 30 days among nests of wild birds in Texas (Bolen, 1967a). It is of interest that the incubation period in this cavity-nesting species seemingly averages somewhat longer than that of the fulvous whistling-duck, a ground-nesting form. Long incubation and fledging periods are typical of cavity-nesting birds, relative to those of ground-nesters.

Fledging period. Cain (1968; 1970) noted that captive-reared ducklings were first observed flying at 53 to 63 days of age. Among pen-raised birds, the age of fledging varied from 56 days in larger and faster-growing ducklings to 63 days for smaller and more slowly growing individuals.

Nest and egg losses. Bolen et al. (1964) reported that of 428 eggs monitored only 83 hatched, a hatching success of 19.4 percent. Predation losses were mainly attributed to raccoons and rat snakes (*Elaphe obsoleta*), but the biggest source of nesting failure was caused by brood parasitism.

In another study, Bolen (1967a) compared the nesting success of natural cavity nests with that of unprotected and antipredator nesting boxes. Of the 32 natural cavity nests, 14 (44 percent) hatched, about the same nesting success rate he found for 13 unprotected boxes. However, 44 protected nesting boxes had a 77 percent nesting success, as compared with a total overall average nesting success of 61 percent for all three types of nesting sites.

There is good evidence for both renesting and double-brooding in this species. Delnicki and Bolen (1976) reported that 19 percent of 57 pairs renested after losing their first clutches, and two pairs renested twice. More significantly, at least one case of double-brooding has been reported (James, Thompson, and Ballard, 2012), making this perhaps the only North American species of waterfowl for which double-brooding has been proven.

Juvenile and adult mortality. There appears to be only a single estimate of mortality rates in this species. Bolen and McCamant (1977) estimated annual mortality rates (both sexes) of 48 to 54 percent, based on both capture-recapture data and band recoveries from hunters. Bolen (1970) commented that although adult sex ratios slightly favored males in his study, there was no statistical indication that females have a higher mortality rate than males.

General Ecology

Food and foraging. One study of the food intake of this species is that of Bolen and Forsyth (1967), which was based on an analysis of 22 stomachs and 11 crops. By volume, these foods consisted of 92 percent plant materials, with a predominance of sorghum grain and Bermuda grass (*Cynodon*) seeds. Later in the summer the seeds of other species, such as smartweeds (*Polygonum*) and water star-grass (*Heteranthera*), were consumed in minor amounts. Virtually no leaves, stems, or roots of any plants were found in the samples. At least locally, rice and corn are consumed in large quantities, and the birds occasionally cause substantial crop damage (Leopold, 1959; Bourne and Osborne, 1978; Bourne, 1981). Animal foods are limited and include gastropod mollusks and various insects.

Unlike the fulvous whistling-duck, this species prefers to forage while standing in shallow water, rather than swimming or diving for its food. Bolen et al. (1964) observed that the birds were rarely seen in water deeper than the length of their legs.

Black-bellied whistling-duck, ducklings

Sociality, densities, territoriality. Like the fulvous whistling-duck, this species is highly social and might be seen in flocks almost throughout the year. It is locally somewhat colonial in nesting. Bolen et al. (1964) estimated a resident population of 250 pairs in the 19,000-acre Lake Corpus Christi near Mathis, Texas, where a large stand of water-killed trees was present. In 1966 26 broods totaling 271 young were seen on the 101-square-mile Laguna Atascosa National Wildlife Refuge, Texas, and by 1975 as many as 380 young had been counted there. It is probable that nesting density is mostly determined by the availability of adequate nest cavities in otherwise suitable habitats. Variations in nesting densities have ranged from as few as 1 to 1.5 nests per square kilometer in Louisiana to more than 300 nests per square kilometer in California, and nests have been found within 50 meters of one another (Hohman and Lee, 2001).

Interspecific relationships. This species and the fulvous whistling-duck often occur in mixed flocks in coastal Mexico but probably have little competition for food or nesting sites. In aggressive disputes, this species typically dominates the smaller fulvous whistling-duck (Cottam and Glazener, 1959). Major nest enemies are probably those that enter nest cavities and destroy the eggs or young, such as raccoons and snakes. In spite of repeated comments asserting this, there is no real evidence that alligators are an important predator on this species. Another hole-nesting species, the muscovy duck, occurs in many of the same habitats, and Bolen (1971) noted that female muscovy ducks sometimes displace nesting black-bellied whistling-duck females.

Daily activities and movements. Like other whistling-ducks, these birds are distinctly crepuscular (Hersloff et al., 1974) to nocturnal in their activities, spending the daylight hours resting or sleeping, and moving out to feeding areas at sundown. No doubt their strong vocalizations are an important means of communication when flying under reduced-light conditions, and the contrasting white upper-wing markings are highly conspicuous in flight. Leopold (1959) mentioned how one's eyes are irresistibly drawn to the flashing wings of these ducks when they are seen in flight.

Social and Sexual Behavior

Flocking behavior. Flock sizes of more than 10,000 birds have been observed in areas of Louisiana that are used for foraging and roosting for fall staging, indicating the highly gregarious tendencies of this species (Hohman and Lee, 2001).

Pair-forming behavior. Virtually nothing has been learned of the details of pair formation in this or any other species of whistling-duck. It must be presumed that the formation of pairs is a very gradual and inconspicuous process, inasmuch as I never observed obvious courtship during two years when this and several other whistling-duck species were under my observation on a nearly daily basis. Perhaps courtship occurs largely or entirely during crepuscular or nocturnal periods.

Copulatory behavior. Copulatory behavior has been described by various writers (Johnsgard, 1965; Meanley and Meanley, 1958). Unlike the fulvous whistling-duck, copulation usually occurs while the pair is standing on shore or in quite shallow water. The male, and sometimes also the female, performs drinking movements not obviously different from those used in normal drinking behavior. Mounting then occurs, and after treading there is a rather inconspicuous postcopulatory display involving mutual calling, but with only slight wing-lifting on the part of the male.

Nesting and brooding behavior. Both sexes apparently participate in nest site selection, and the male also assists with incubation. No down is plucked from the breast of either sex during incubation, although both sexes develop vascularized incubation patches. Even without down in the nest, quite possibly the heat of summer is responsible for some embryonic development (Cain, 1970). When the young hatch, both sexes carefully tend them. Typically, one adult swims in front of and the other behind the brood. When threatened by a predator, one parent often leaves the group to decoy and harass the intruder, while the other leads the brood toward safety. Young have sometimes been observed riding on the backs of swimming adults (Bolen et al., 1964), an activity apparently not reported from any other whistling-ducks but present in several swans and some sea ducks (*Bucephala*, *Mergus*).

Postbreeding behavior. Little definite information is available on postbreeding behavior, but the Texas population apparently begins its southward migration not long after the young have grown and the adults have completed their postnuptial molts. In Texas this departure usually occurs between August and October, although some migrants might remain into November.

Tribe Aythyini (Pochards)

Pochards comprise a readily recognizable group of about 12 mostly medium-sized diving ducks that have been variously described as "inland divers," "freshwater divers," or "bay diving ducks." The less familiar "pochard" term is an English word dating back to the mid-sixteenth century. It is of uncertain etymological origin but provides a convenient one-word descriptor. Possibly it was derived from "poacher," a term sometimes used for dabbling ducks such as wigeons that try to steal food from diving species when they return to the surface.

In the British Isles the term "pochard" refers only to the Eurasian pochard (*Aythya ferina*) and the red-crested pochard (*Netta rufina*). However, as Delacour (1959) has done, the name is here used taxonomically to include both of the genera *Aythya* (the broad-billed pochards) and *Netta* (the narrow-billed pochards).

Pochards differ from their close relatives the typical surface-feeding ducks (*Anas*) in several anatomical respects, although some transitional forms (*Marmaronetta, Rhodonessa*) exist, as I discovered on the basis of various morphological characters (Johnsgard, 1961a, 1961b). The legs of pochards are situated somewhat farther back on the body than is true of *Anas*, so pochards are less adept at walking on land and making leaping takeoffs impossible. Their feet and associated webs are also larger relative to body mass, and they have notably longer outer toes, probably increasing diving efficiency but not making leaping takeoffs possible. They are mostly temperate zone, northern hemisphere species, and most are primarily vegetarians. Pochards are strong, swift fliers, and many undertake long seasonal migrations. However, one tropical Indian species (the extinct pink-headed duck, *Rhodonessa caryophyllacea*) was apparently nonmigratory.

No iridescent speculum pattern is present on the secondary wing feathers of pochards, but in many species the secondaries are contrastingly white, or at least distinctly paler than the rest of the wing. Additionally the heads of males of some species (especially scaup) exhibit iridescent coloration. Like other true ducks, pochards molt their body feathers twice a year, and males of the more sexually dimorphic species alternate between relatively distinctive plumage colors during the pair-forming and early breeding season and more female-like (basic) plumages for the rest of the year.

Like male dabbling ducks, adult males of all pochard species have tracheal tubes that are variably inflated at the junction of the two bronchi, forming a hollow structure, the tracheal bulla. However, in contrast to the rounded and entirely bony structure of the tracheal bulla in male dabbling ducks, the structure in male pochards is angular and partially membranaceous. Interestingly, tracheal bullae of transitional structures between those of dabbling ducks and pochards are present in the marbled teal (*Marmaronetta*) and pink-headed duck (*Rhodonessa*) (Johnsgard, 1961a, 1961b, 1961c).

A tracheal bulla is lacking in female pochards, but vocalizations are generated by a syrinx. Located at the junction of the bronchi and trachea, the syrinx has paired vibratory membranes and associated muscles that control membrane tension. The syrinx generates sounds when air is passed over its paired membranes, with the bird's vocal flexibility determined by the syrinx's ability to modify vocal amplitude, pitch, and harmonic content. Like those of most ducks and geese, pochard vocalizations tend to be low in pitch and rich in harmonic content, facilitating acoustic variability and increasing the potential for achieving individual sound recognition.

The associated calls of male pochards are soft, species-specific vocalizations that lack whistled elements and have relatively little volume and carrying power. They are usually uttered in association with visually distinctive neck and head postures used during pair formation, such as "head-throws," "sneaks," and "kinked-neck" displays. Like many other ducks, female pochards take active roles in pair-forming behavior, primarily by "inciting" a prospective mate to threaten or attack potential rivals, and probably thus being able to identify and select the fittest mate.

The bills of pochards vary from being relatively long, narrow, high-based, and adapted for grazing on the stems and roots of underwater vegetation, as is true of canvasbacks, to the wider and flatter bills of scaups, which are much more dependent on bottom-dwelling invertebrates, to those of redheads and ring-necked ducks, which have transitional forms and foraging ecologies (Hughes and Green, 2005). Thus, depending on the species, the predominant food might be crustaceans and other invertebrates, vegetative materials, or both. Foraging usually occurs at depths of less than ten feet but sometimes is performed in somewhat deeper waters.

The usual breeding season habitats of pochards are shallow, temperate-zone freshwater marshes, edged by many emergent plants such as rushes and reeds, among and on which nesting occurs. Pochard breeding marshes are usually interspersed with areas of open water providing for easy surface movement and enough space to take flight by running along the water surface until flying speed is attained. In flight, pochards fly swiftly without rapid changes in direction, in generally loose formations. Perhaps second only to mergansers, species such as the canvasback are among the swiftest flyers of all waterfowl; a chased canvasback once was clocked in excess of 70 miles per hour.

All pochards become sexually mature and most probably breed during their first year, although yearling females are less likely to breed than males. Pochards form temporary monogamous pair-bonds that persist for a single breeding cycle but normally terminate during the female's incubation period. Pochards usually nest very close to or over water; in many species the nest is built above the water surface, on reed mats or similar floating vegetation. Incubation and brood-rearing are performed by females only. Fledging periods among pochards average about 20 percent longer than those of comparably-sized dabbling ducks (e.g., mallards at 55–60 days vs. canvasbacks at 70–75 days), which might influence northern range limits for breeding or renesting in some pochard species.

In some pochard species a substantial percentage of females also engage to varying degrees in brood-parasitism ("dump-nesting") behavior, depositing some of their eggs in the nests of conspecifics, or even in those of other marsh-nesting birds. Usually only a small percentage of these alien eggs hatch into young that survive to fledging, but perhaps enough do so to at least make this a viable supplemental breeding strategy for some species. Renesting following nest failures is typical of many species, but raising two broods within a single breeding season has not been observed in any of the pochards.

It is common for female pochards to abandon their broods somewhat prior to fledging. Females then begin their postbreeding molt, with its associated period of flight feather replacement and several weeks of temporary flightlessness, usually several weeks after the males have undergone their flightless periods. Limited molt migrations are typical of some pochards, during which postbreeding males and failed or nonbreeding females migrate prior to molting to locations where food supplies and habitat characteristics are favorable for spending this stressful and vulnerable period. Several pochard species undertake extensive molt migrations to favorable molting areas.

North America has five widely distributed species of native pochards, one of which (the greater scaup) also occurs widely in the Old World. Additionally, hundreds, if not thousands, of North American records of the Eurasian tufted duck now exist, so an abbreviated descriptive summary of that species is included. Since one other Old World species, the Eurasian pochard, has rarely been observed in the Aleutian Islands and very rarely reported from the continental mainland, an abbreviated description of that species is also included.

Canvasback
Aythya valisineria (Wilson) 1814

Other vernacular names. Canvas-backed duck, can

Range. Breeds from central Alaska south though British Columbia, eastern Washington, and southeastern Oregon to north-central California and east to southeastern Idaho, northeastern Utah, eastern Wyoming, northern Colorado, northern Nebraska, the Dakotas, and western Minnesota. Winters from southern Canada south along the Atlantic and Pacific coasts (especially in central California), the Gulf coast, lower Mississippi River valley, and Atlantic coast from Massachusetts south to Florida, and less frequently in the continental interior south to central Mexico.

Subspecies. None recognized.

Measurements. *Folded wing:* Delacour, 1959: Males 225–242 mm; females 220–230 mm.
   Culmen: Delacour, 1959: Males 55–63 mm; females 54–60 mm.

Weights (mass). Nelson and Martin (1953): 62 males, ave. 2.8 lb. (1,268 g); 79 females, ave. 2.6 lb. (1,178 g). Combined fall data of Bellrose and Hawkins (1947) and Jahn and Hunt (1964): 8 adult males, ave. 2.99 lb. (1,356 g); 14 immature males, ave. 2.83 lb. (1,283 g); 5 adult females, ave. 2.49 lb. (1,129 g); 9 immature females, ave. 2.47 lb. (1,120 g). Nelson and Martin (1953): max. male weight 3.5 lb. (1,577 g); max. female weight 3.4 lb. (1,542 g).

Identification

In the hand. Canvasbacks are the only North American pochards that have a culmen length in excess of 50 mm (or 2 inches); additionally the bill is uniquely sloping from its base to the tip and lacks a pale band near the tip. Supplementary criteria include the presence of vermiculated upper wing-coverts, with the white predominating over the dark rather than the darker tones predominating.

In the field. When on the water, *male* canvasbacks appear to be nearly white on the mantle and sides, whereas male redheads are distinctly medium gray, and the longer, more sloping head of the canvasback is usually evident. Compared to the redhead, the head is a duller chestnut brown, darker above and in front of the red eyes; in redheads the head is a coppery red and little if at all darker in front of the yellow eyes. *Female* canvasbacks are distinctly longer-bodied than female redheads and lighter in brownish tones, with the brown breast usually distinctly darker than the more grayish sides, whereas in redheads the difference in color between the breast and the flanks is not very apparent. Both sexes appear longer necked than redheads; in males this is accentuated by the extension of the reddish brown color beyond the base of the neck.

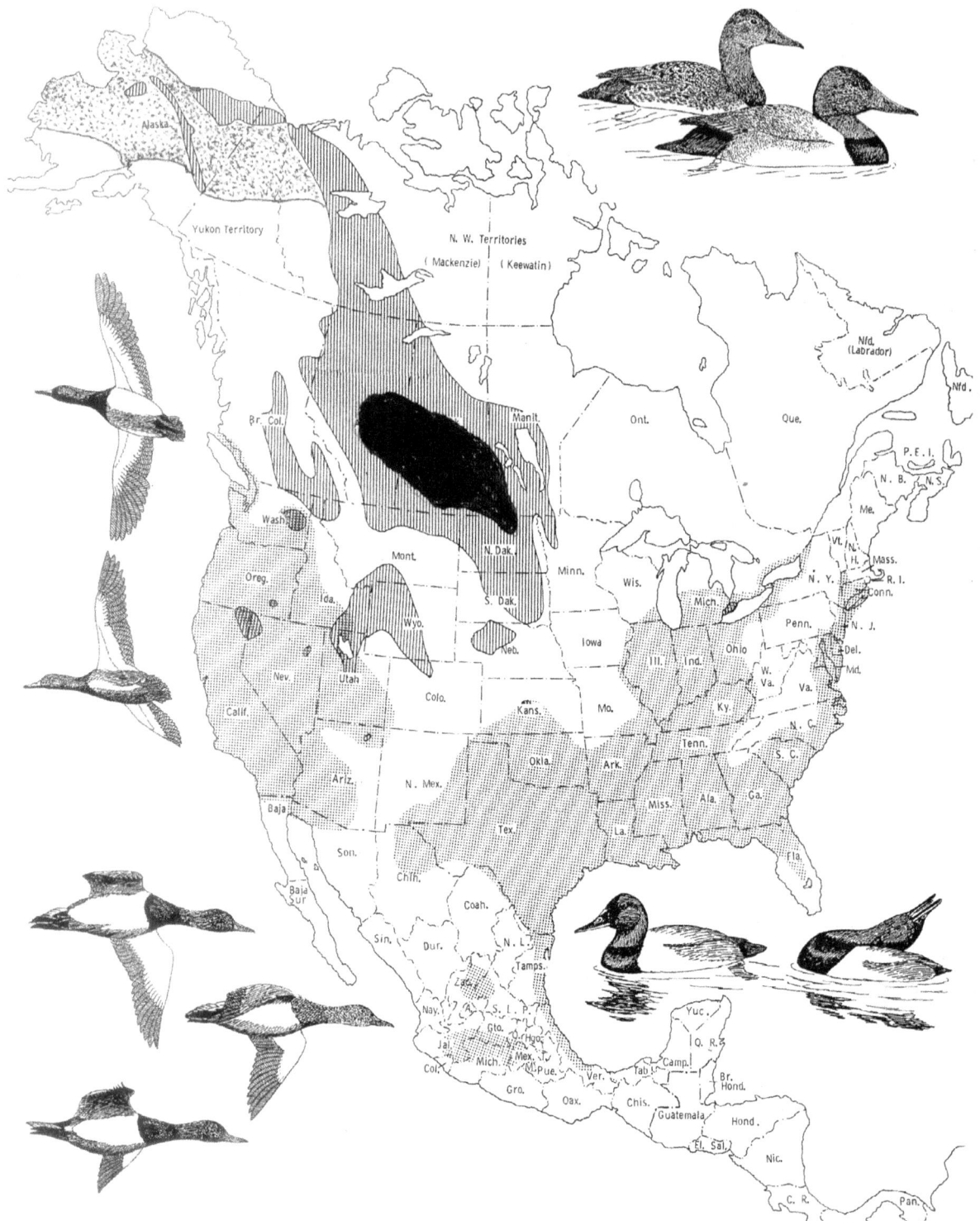

The breeding (hatched, with denser concentrations inked), wintering (shaded), and marginal (stippled) range of the canvasback.

In flight, this difference is also apparent; the black breast of the male canvasback is more restricted and does not reach the leading edge of the wings, whereas in redheads the black breast extends to the front of the wings. In females the brownish breast appears sharply separated from the pale grayish sides, but in female redheads the brown breast color is continuous with the brown of the sides and flanks. Except during courtship, canvasbacks are relatively quiet; the male's cooing courtship call (uttered only on the water) might be heard frequently during spring.

Age and Sex Criteria

Sex determination. A reddish eye color indicates a male in any adult plumage, as does the presence of rusty brown on the head or black feathers on the breast or tail coverts. However, since females are extensively vermiculated, this trait is not diagnostic for sex. Even in full eclipse the head of the male is relatively dark and lacks the pale areas around the eyes and the pale throat typical of females. Dzubin (1959) noted that by 30 days of age males begin to exhibit lighter scapular feathers than do females.

Age determination. Juveniles are female-like. Immature birds of both sexes might still carry some juvenal tertials, which are usually frayed to a pointed tip and are iron gray, with or without white flecking, whereas in adults they are rounded and always have some vermiculations of flecking. The presence of any juvenal tertial coverts, middle coverts, or greater coverts indicates immaturity and can be easily recognized by their more uniformly grayish and unflecked or lightly flecked pattern, compared with the vermiculated first-winter or adult feathers (Carney, 1964).

Distribution and Habitat

Breeding distribution and habitat. The canvasback occupies a breeding range and habitat comparable to that of its close European and Asian relative, the common pochard. It tends to have a somewhat more northerly distribution than that of the redhead, although the habitat requirements of these two species are quite similar. In Alaska the canvasback has a relatively wide breeding distribution and is a common summer resident over much of that state (Hanson, 1960). Its northernmost known occurrence is north of the Arctic Circle.

In Canada the canvasback ranges from northern Yukon Territory and the Anderson River delta of the Northwest Territories southeastward to central Ontario, including central and southern British Columbia and especially the "prairie pothole" region of Alberta, Saskatchewan, and Manitoba. Breeding bird surveys in 2010 estimated 585,000 birds were within the "Traditional" aerial survey route extending from Alaska and northwestern Canada to the northern and central plains of Canada and the northern plains of the Dakotas and Montana. Regionally, 33 percent were found in southern Saskatchewan; 12 percent in the Northwest Territories, British Columbia, and northern Alberta; 10 percent in the Alaska-Yukon region; and the remainder in southern Alberta, southern Manitoba, the Dakotas, and Montana (Baldassarre, 2014).

The breeding range in the United States south of Canada is now highly disrupted and probably declining because of the extensive marsh destruction and drainage that has occurred in the prime areas of the canvasback's range. In eastern Washington the canvasback is a rare nesting bird in Adams and Lincoln Counties (Yocom, 1951). In Oregon an estimated 500 pairs nest at Malheur National Wildlife Refuge, and it is uncommon to fairly common in the wetlands of Harney, Lake, and Klamath Counties (Gilligan et al., 1994), south to the Klamath Lake–Tule Lake region of southern Oregon and adjacent northern California. It also nests locally in the Ruby Lake area of Nevada and in northern Utah, northern Arizona, southern Idaho, northern Colorado, and eastern Wyoming.

The heart of its United States nesting range is probably in the prairie pothole area of eastern Montana and the Dakotas as well as the Sandhills lakes of Nebraska. The southern limit of regular breeding in the prairie states is apparently Kansas (Johnstone, 1964), where it has bred at Quivira National Wildlife Refuge and Cheyenne Bottoms Waterfowl Area (Thompson et al., 2011). To the east, the canvasback nests locally in northern Minnesota (Lee et al., 1964), and has rarely nested in Wisconsin (Jahn and Hunt, 1964). There are a few breeding records for Michigan (Brewer, McPeek, and Adams, 1991) and New York (Mowbray, 2002).

The preferred breeding habitat of canvasbacks consists of shallow prairie marshes surrounded by cattails, bulrushes, and similar emergent vegetation, large enough and with enough open water for easy takeoffs and landings, and with little if any wooded vegetation around the shoreline. Dwyer (1970) noted a much higher breeding canvasback population outside than inside Riding Mountain National Park in Manitoba, apparently because of the reduced numbers of trees around the breeding ponds. Keith (1961) found the highest use of canvasback pairs per unit of shoreline on a shallow lake with a maximum depth of eight feet, having scattered strands of bulrushes, shorelines dominated by rushes (*Juncus*), sedges (*Carex*), and spike rush (*Eleocharis*), and several cattail-covered islands. Brood use per acre of water was also highest on this lake; female canvasbacks apparently moved from smaller nesting marshes to larger wetlands following hatching. Hochbaum (1944) noted that canvasbacks tend to use larger bays in the Delta, Manitoba, region than do other resident diving ducks, which frequent sloughs (shallow marshes) and potholes to a greater extent.

Population. Breeding bird surveys by the US Fish and Wildlife Service and Canadian Wildlife Service estimated national surveyed canvasback populations from 1955 to 2015 to average slightly over 500,000 birds. A relatively stable population prevailed over most of this time except during periodic droughts, and a slight upward trend has occurred since the early 2000s (Zimpfer et al., 2015). The national 2016 population was estimated at 700,000 birds (US Fish and Wildlife Service, 2016). The estimated US hunter kill during the 2013 and 2014 seasons averaged about 130,000 birds (Raftovich, Chandler, and Wilkins, 2015). The Canadian kill has averaged less than 10,000 birds in recent years, and the species has occasionally been listed as a Species of Concern in Canada during the population declines of the 1980s.

Wintering distribution and habitat. To a rather surprising degree, the interior-nesting canvasbacks tend to move to coastal areas for the winter months. On the Pacific coast some wintering occurs as far north as southern British Columbia and the Puget Sound region of Washington. Some wintering also occurs in

Fig. 9. Canvasback, adult male swimming

western Oregon, but the center of the canvasback's Pacific coast wintering habitat is the San Pablo Bay of central California.

Recent winter surveys by the US Fish and Wildlife Service indicate that about one-fourth of the continental canvasback population winters in the Pacific Flyway, most of it north of the Mexican border. During the 2000–10 winter surveys, more than 50 percent of the estimated total of 300,000 birds were found in the Mississippi Flyway, 20 percent were in each of the Atlantic and Pacific Flyways, and less than 10 percent were in the Central Flyway (Baldassarre, 2014). In Mexico, the canvasback is a relatively minor component of the wintering waterfowl, with the largest numbers found on the Pacific coast and in the interior. Leopold (1959) noted that during 1952 surveys most of the canvasbacks seen were on Lakes Chapala and Patzcuaro, with the remainder primarily found near Tampico, Veracruz, whereas in more recent surveys two-thirds of the birds have been found on the lagoons of coastal Tamaulipas, especially Laguna Madre, and with smaller numbers on the Tamiahua Lagoon, Veracruz. Small numbers also winter in the Mexican interior and along the Pacific coast.

In the Atlantic Flyway, which harbors the majority of the North American canvasback population, wintering birds commonly occur as far south as central Florida (Chamberlain, 1960) and north to coastal New England but concentrate in the Chesapeake Bay region. This region typically supported nearly three-fourths of the Atlantic Flyway canvasback population during the 1950s, or almost half of the entire continental population (Stewart et al., 1958), and in the early 2000s the Chesapeake Bay region still supported more than 80 percent of the birds wintering in the Atlantic Flyway (Baldassarre, 2014).

The Detroit River–Lake St. Claire area and the coastal area of the Mississippi River valley represent other traditional major wintering locations in the eastern United States, and the San Francisco Bay region is the most important wintering area in the Pacific Flyway. Since the 1960s the relative number of canvasbacks wintering in the Mississippi Flyway has increased to about 20 percent of the US total, and the Atlantic and Pacific Flyways have decreased somewhat in their relative importance.

Stewart (1962) observed that the optimum canvasback habitat in the Chesapeake Bay area consisted of fresh and brackish estuarine bays containing extensive beds of submerged plants or abundant invertebrates, especially certain thin-shelled clams and small crabs. Beds of wild celery (*Vallisneria*) in freshwater estuarine bays are heavily utilized by canvasbacks, as are pondweed (*Potamogeton*), wigeon grass (*Ruppia*), and eelgrass (*Zostera*) in the brackish bays. Brackish estuarine bays are the principal wintering habitats, with both saltwater and freshwater estuarine bays being used relatively little.

General Biology

Age at maturity. Canvasbacks probably normally reproduce when a year old but in captivity are particularly difficult to breed successfully. Ferguson (1966) noted that only one of 14 aviculturists reported breeding by yearling canvasbacks, and most indicated that initial breeding occurred in the second or third year. Hochbaum (1944) also noted that captive canvasbacks that bred at Delta, Manitoba, were all more than a year old, but he believed that wild canvasback females commonly nest when a year old, and that males are also physically able to reproduce at that age, although not all might be successful.

Pair-bond pattern. Pairs are reformed each winter and spring during a prolonged courtship period. Weller (1965) found that up to 10 percent of the female canvasbacks he observed between December and March were paired, whereas 41 percent were paired during March and April counts. Smith (1946) observed intense pair-forming activities during mid-April in Minnesota. Hochbaum (1944) observed that most canvasbacks were not paired on their arrival in southern Canada, but pair formation reaches a peak in late April and early May, and most birds are paired after the middle of May.

Nest location. Lee et al. (1964b) reported that canvasbacks in Minnesota nested over water in emergent vegetation that ranged from 14 to 48 inches high and averaged 34 inches, higher than the averages they found for both ring-necked duck and redhead. Seventeen nest sites averaged 11 yards from open water and ranged from 0 to 55 yards. Preference was shown among canvasbacks for nesting in smaller bulrush-dominated marshes having some open water present.

Fig. 10. Canvasback, pair resting

Stoudt (1971) found that 80 percent of the 172 canvasback nests he found in Manitoba were in cattail cover, and similar preferences for cattail have been reported by Smith (1971) in Minnesota and Keith (1971) in Alberta. Hochbaum (1944) noted a strong preference for nesting in hardstem bulrush (*Scirpus acutus*) at Delta, Manitoba, with cattails and reed (*Phragmites*) also being accepted, but softstem bulrush (*Scirpus validus*) was not used as a nest cover. Townsend (1966) found a high usage of reeds and a low usage of sedges for canvasbacks, just the opposite of the situation for the ring-necked duck and lesser scaup. Furthermore, canvasbacks typically placed their nests closer to large areas (over 50 × 50 feet) of open water than did those species, and all the canvasback nests found were within 40 feet of such areas of water.

Clutch size. Smith (1971) determined an average clutch size of 7.4 eggs for 118 Minnesota nests. In a 12-year study, Stoudt (1982) reported that 497 completed and nonparasitized canvasback nests had an average clutch of 8.2 eggs, but average clutch sizes declined gradually in later clutches. Erickson (1948) found that 15 nonparasitized canvasback nests had 9.9 eggs present during initial nesting efforts, compared to an average clutch of 8.6 eggs in nonparasitized renesting attempts. Among 74 parasitized nests, there was an average of 7.0 host eggs and 6.1 parasitic eggs. Kruse et al. (2003) reported an average of 6.4 eggs for 223 nonparasitized nests at Ruby Lake National Wildlife Refuge, Nevada, with 155 successful nests averaging 7.0 eggs; this location is relatively far south for canvasback breeding, although they have bred at Bosque del Apache National Wildlife Refuge in southern New Mexico.

Incubation period. Hochbaum (1944) noted that although ranges in incubation periods of 23 to 29 days had been estimated, most eggs hatched after 24 days under artificial incubation conditions at Delta, Manitoba.

Fledging period. Fledging occurred 56 to 68 days after hatching among eight hand-reared birds (Dzubin, 1959). Lightbody and Ackney (1984) reported a 71-day fledging period for eight hand-reared birds.

Nest and egg losses. Sowls (1948) noted a 48 percent hatching success for 24 nests, and Lee et al. (1964b) recorded a 25 percent hatching rate for 16 nests, with predators accounting for half the losses and the striped skunk being the primary egg predator. Smith (1971) and Stoudt (1971) estimated hatching rates of 48 and 65 percent, respectively, with striped skunks, American crows, and black-billed magpies as probable predators. Crows also accounted for many of the nest losses in Sowls's (1948) study.

Parasitically laid eggs in canvasback nests include both interspecific parasitism, mostly involving redheads, and intraspecific parasitism. Hochbaum (1944) found that redheads parasitized 58 percent of 38 canvasback nests, with an average of 6.4 redhead eggs deposited per nest (but only 5 percent of 56 redhead clutches contained any canvasback eggs). Erickson (1948) found that this parasitism adversely affected nesting success, with 91 percent of the eggs hatching in unparasitized nests that he studied, compared to 77 percent of the eggs in those that were parasitized. Likewise, a lower percentage of nests hatched when parasitically laid eggs were present, and a smaller average number of canvasback young hatched per nest.

Sorenson (1997) found that among 179 canvasback nests in Manitoba, redheads parasitized 80 percent of the hosts' nests, typically displacing 0.31 host egg for every parasitic egg laid, thereby reducing the number of host egg hatching and lowering brood survival rates (87 to 73 percent). In another larger, multiyear study, Stout (1982) reported that redheads parasitized 57 percent of 1,131 canvasback clutches, a result similar to a rate of 63 percent among 134 canvasback nests in another study (Serie, Trauger, and Austin, 1992). These authors found no differences in nest success between parasitized and nonparasitized nests but observed a slightly lower hatching success for canvasback eggs in the two groups: 87 percent vs. 77 percent, respectively. In a Nevada study, redheads parasitized 70 percent of 811 canvasback nests, but hatching success of canvasback eggs was not significantly affected unless at least 4 parasitic eggs were present in the nest (Kruse et al., 2003).

Intraspecific brood parasitism also can have deleterious effects on canvasback breeding. Sorenson (1993) found that such parasitism affected 36 to 41 percent of 179 clutches of canvasbacks in Manitoba, accounting

Canvasback, breeding males and females

for 9.7 to 13.3 percent of the total canvasback eggs in the sample. The resulting maximum hatching success of host eggs was 79 percent, as compared with 29 percent for parasitically laid eggs.

Following clutch loss, renesting efforts are rarely made by canvasbacks; Stout (1982) identified only 12 renesting females out of almost 1,900 nests studied. Renesting probabilities evidently depend on the female's age, the time of clutch loss, and environmental conditions, with most renesting attempts made by females older than two years of age, those that lose their first clutch early in the breeding seasons, and during years of nondrought conditions (Doty, Trauger, and Serie, 1984).

Juvenile mortality. Smith (1971) and Stoudt (1971) both estimated rearing success rates of about 80 percent. Geis (1959) similarly judged that an average of 77.4 percent of canvasback pairs were successful in raising broods and that an average of 5.8 ducklings per brood fledged. However, frequent brood disruption and mergers of unrelated broods make brood size counts of older ducklings unreliable as estimates of duckling mortality.

Based on the banding of unfledged canvasbacks, a high first-year mortality rate of 77 percent has been estimated (Geis, 1959). Along with the serious effects of drought on nesting success, the associated devastating effects on the specialized nest-site requirements of canvasbacks, and the species' often high juvenile mortality rates, are major reasons for the periodic population declines of canvasbacks.

Studies by Anderson et al. (2001) indicated that survival of juveniles was positively correlated with body mass and negatively correlated with hatching date, so underweight ducklings hatched late in the season has only a 9 percent survival rate, whereas those with high body mass and that had hatched early in the breeding season had survival rates as high as 75.1 percent.

Adult mortality. Geis (1959) estimated that an annual mortality rate of 35 to 50 percent is typical of canvasbacks after their first year of life. Boyd (1962) calculated an average 41 percent annual mortality rate based on these same data. Females have considerably higher mortality rates than do males, which at least in part accounts for the seriously unbalanced sex ratios that have generally been observed for canvasbacks (Olson, 1965).

Other more recent estimates of annual survival dates include Anderson et al. (1997), who calculated a 66.8 to 69.9 percent rate for adult females and a 40.4 to 41.4 rate for juveniles. Nichols and Haramis (1980a) estimated an adult male survival rate of 72.6 to 81.8 percent and 56.1 to 69.0 percent for females, based on 2,600 recoveries.

General Ecology

Food and foraging. The attraction of canvasbacks to wild celery beds in the northeastern states is very well known, and in that area they utilize both the seeds and vegetative parts of this plant extensively. Pondweeds (*Potamogeton*) play a secondary role there, but in the western states and the southeast their vegetative parts and seeds largely replace wild celery as the primary food. The vegetative parts of arrowhead (*Sagittaria*) and banana water lily (*Nymphaea flava*) are also of importance in the southeastern states (Martin et al., 1951).

In Minnesota, canvasbacks have traditionally been attracted to Lake Christina, which is large and shallow and has abundant growths of sago pondweed, wigeon grass, and naiad (*Najas*), of which the sago pondweed is selectively consumed by canvasbacks (Smith, 1946). Cottam (1939) also determined that pondweeds are the most important food for both canvasbacks and redheads. Some captive immature canvasbacks were found to consume from 2 to 3 percent of their body weight per day in natural foods, or an average of 0.78 pound of wet-weight materials per day (Longcore and Cornwell, 1964).

Stewart's (1962) study of canvasbacks shot in the Chesapeake Bay area indicated that various mollusks and crustaceans, especially macoma bivalves (*Macoma*) and mud crabs (*Xanthidae*), are important foods for birds wintering in brackish estuaries and the Patuxent River.

Sociality, densities, territoriality. Few quantitative data on canvasback breeding densities are available. Lee et al. (1964b) noted that in a 2.5-square-mile study area of Mahnomen County, Minnesota, 2.5 to 7.0 pairs were present per square mile over a four-year period. Keith (1961) found an average of 2 pairs occupying

Canvasback, breeding pair swimming

183 acres of impoundments during five years of study in Alberta, the equivalent of about 7 pairs per square mile of wetlands.

Dzubin (1955) noted that canvasbacks composed up 10 percent of the breeding ducks in an area of southern Manitoba having 97.8 breeding duck pairs per square mile, or about 10 canvasbacks per square mile. Stoudt (1969) observed that the peak densities of canvasbacks on five prairie study areas in Saskatchewan, Manitoba, and South Dakota ranged from less than 1 to 11 pairs per square mile.

Hochbaum (1944) believed that territorial boundaries in canvasbacks and other pochards are less rigid than in surface-feeding ducks, and he never observed direct attacks associated with apparent territoriality. He did, however, believe that spacing of breeding pairs exists in this species. Dzubin (1955) noted that canvasbacks were highly mobile during the prelaying and incubation phases of reproduction, and that certain areas had overlapping usage by different pairs, so that the concept of a home range, rather than a classic territory, seemed more appropriate in describing this species.

Interspecific relationships. Perhaps because of the similarities in nest site preferences, the canvasback is conspicuously affected by the parasitic nesting tendencies of redheads (Weller, 1959). Canvasbacks not only socially parasitize other females of their own species (see the section "Nest and egg losses") but also have been known to lay their eggs in the nests of both redheads and ruddy ducks.

Skunks, crows, and raccoons (and no doubt a large number of other predators and scavengers) have been found to be responsible for high losses of eggs and ducklings, but the vulnerable status of canvasback populations is more directly related to human activities: destruction of breeding habitat, pollution or other degradation of critical wintering areas, and possible overshooting of females. Female losses are especially serious, since females are much more vulnerable than males to shooting, and they represent a limiting factor in potential production because of the distorted sex ratio typically present among adults, in which males variably outnumber females.

General activity patterns and movements. Hochbaum (1944) has provided an excellent account of the daily and seasonal activities of canvasbacks on their nesting grounds.

Dzubin (1955) observed that a male canvasback occupied a home range with a maximum length of 3,900 yards during the breeding season, and that his mate was somewhat less mobile, so that an overall home range of about 1,300 acres was estimated. Male canvasbacks apparently do not defend any of their home range but show aggression when other males approach their mates.

To a greater extent than is apparent with most ducks, canvasbacks appear to migrate in seasonal "waves," with the dates of arrival both in spring and fall being fairly predictable (Smith, 1946; Jahn and Hunt, 1964). In spring, paired birds typically reach their breeding grounds first, followed later by unpaired flocks. There apparently is a differential migration of ages and sexes during the fall flights, but differential sex and age vulnerability to hunting confuses the picture in interpreting fall movements.

Social and Sexual Behavior

Flocking behavior. Hochbaum (1944) noted that during spring, arriving migrant canvasbacks are in small flocks that usually number about 4 to 12 birds and rarely exceed 20. On the other hand, fall groups are typically quite large and gain in size as they move southward. Large concentrations are facilitated by the usually restricted number of favored feeding areas. Smith (1946) reported that on the 4,000-acre Lake Christina in Minnesota, maximum concentrations of about 30,000 birds were counted during the spring migration period. He noted that it was not unusual to see a flock of several thousand birds about a hundred yards off-shore engaged in courtship activities.

Pair-forming behavior. The pair-forming behavior of canvasbacks (Fig. 11) has been well described by Hochbaum (1944). His account, as well as observations by Smith (1946) and Weller (1965), indicates that pair-forming activities begin in late winter and reach their peak in mid-April, during late stages of spring migration and arrival on the breeding areas.

Fig. 11. Sexual behavior of canvasback, including (A) head-throw, (B) sneak, (C) neck-stretching, (D–F) kinked-neck call, and (G–H) postcopulatory bill-down posture (after Johnsgard, 1965).

Pair-forming displays of the canvasback, as first described by Hochbaum, have provided the basic terminology for the displays of all pochard species. A courtship "kinked-neck" call (Fig. 11D–F) is common. The same call is uttered during a head-throw (Fig. 11A). There is also neck-stretching (Fig. 11C), a crouching "sneak" posture (Fig. 11B), and a threat-like attitude. Females perform inciting with strong neck-stretching, and inciting occurs in the same situations as with surface-feeding ducks. Wing-preening displays have not been observed in canvasbacks, but preening of the dorsal region is a major precopulatory display of all pochard species (Johnsgard, 1965). Aerial chases of the female, as described by Hochbaum, do occur frequently in canvasbacks, but whether the aerial tail-pulling he described is a typical aspect of pair formation or rather related to attempted rape behavior is still somewhat uncertain.

Copulatory behavior. Copulation in canvasbacks is normally initiated by the male performing alternate bill-dipping and dorsal-preening movements. These are not highly stereotyped displays and might be overlooked by a casual observer. The female might perform the same displays, but she commonly assumes a prone posture on the water without prior displays. Treading lasts several seconds, and as the male releases the female's nape, he typically utters a single courtship call, then swims away in a rather rigid bent-neck posture with his bill pointed nearly vertically downward (Fig. 11G–H). The female usually begins to bathe immediately (Johnsgard, 1965).

Nesting and brooding behavior. Female canvasbacks typically spend a considerable period searching for suitable nest sites and might abandon one or two nests before settling on a final location. The first eggs might be laid before the nest is completed, and might be "dropped" in various places, sometimes in other nests. Eggs are laid in the morning, usually shortly after sunrise, at the rate of one per day. Down is often initially placed in the nest after the third or fourth egg, and is usually quite abundant by the time the clutch is completed. The female might be on the nest nearly continuously while the last two eggs are being deposited, and apparently begins incubation with the laying of the last egg.

During incubation the female sometimes takes short rest periods off the nest during morning and evening hours, but these are reduced as incubation proceeds. The period between initial pipping and hatching varies from 18 to 48 hours (Hochbaum, 1944).

Following hatching, the female takes her brood from the nest site to the open water of larger ponds and shallow lakes, feeding heavily in the morning and evening but sometimes also at midday. The hen typically does not defend her young as intensively as do female surface-feeding ducks and usually abandons them before they have fledged and when she begins to undergo her postnuptial molt (Hochbaum, 1944).

Postbreeding behavior. Although the male accompanies the hen while she is searching for nest sites, he spends much of his time at a regular loafing site once the nest site is chosen. As soon as the clutch is completed, he typically deserts his mate (Hochbaum, 1944), although he might also remain associated with her until about mid-incubation (Dzubin, 1955). Thereafter he starts to associate with other males in similar reproductive condition and begins his postnuptial molt. Molting often occurs after undertaking a premolt migration to large lakes that provide both shallow water and abundant aquatic vegetation, such as pondweeds (Baldassarre, 2014).

Eurasian (Common) Pochard
Aythya ferina (Linnaeus) 1758

Other vernacular names. European pochard, pochard

Range. Breeds in the British Isles, southern Scandinavia, and from western Russia east through western Siberia to Lake Baikal; also south to the Netherlands, Germany, Romania, and east past the Black and Caspian Seas and the steppes of southern Russia to the Yarkand River in western China. Winters from its breeding range south to northern Africa and east to the Persian Gulf, India, Myanmar (Burma), southern China, and Japan.

Subspecies and range. No subspecies recognized.

Measurements. *Folded wing:* Delacour (1959): Males, 207–224 mm; females, 201–212 mm. Owen (1977): Adult males, ave. 215.1 mm; adult females, ave. 208.5 mm.
    Culmen: Delacour (1959): Males, 45–51 mm; females, 43–47 mm. Owen (1977): Adult males, ave. 47.0 mm; adult females ave. 45.0 mm.

Weights (mass). Dementiev and Gladkov (1967): Males (September) 930–1,100 g, ave. 998 g; females, 900–995 g, ave. 947 g. Owen (1977): Adult males, ave. 854 g; adult females, ave. 809 g.

Identification and Field Marks

Males in breeding plumage have a uniformly ruddy chestnut head and neck and a black breast. The back, scapulars, flanks, and underparts are predominantly white, with black vermiculations of varying coarseness except on the abdomen. The lower back, rump, and tail coverts are black, and the tail is nearly black. The upper wing-coverts are a vermiculated gray, and the secondaries are also gray, with no obvious speculum, whereas the primaries are more brownish, especially toward their tips. The iris is red, the bill is mostly dark bluish, with a pale band between the nostrils and a blackish tip. The legs and feet are bluish gray with darker webs. Males in postbreeding (basic) plumage resemble females but retain a yellow to reddish iris and have some vermiculations on the upperpart feathers.

Females have a brownish head, which is darker on the crown and hindneck and grades to buff on the chin and throat. There is also a buffy eye-ring and a vague pale stripe behind the eye. The upperparts are brown, or gray vermiculated with brown, whereas the rump, tail coverts, and tail are brownish black. The breast is brown, shading to pale gray on the abdomen and flanks. The upper wing-coverts are grayish brown, without vermiculations; the secondaries are gray; and the primaries are dark brown. The iris is brown, whereas

Fig. 12. Eurasian pochard, two views of adult male

the bill, legs, and feet are similar in color and pattern to those of the male. *Juveniles* of both sexes closely resemble adult females but have more mottled underparts.

In the highly unlikely event that Eurasian pochards are seen in company with canvasbacks or redheads, the broad and conspicuous pale band across the bill is the best distinguishing mark; otherwise the birds look almost perfectly intermediate between these two species and might be regarded as possible hybrids. Females have the usual two general pochard calls: a growling *brerr* or *errr* associated with inciting and an

Eurasian pochard, breeding pair resting

aggressive *pack* or *back*. Males utter a soft breathing *wiwierrr* and a louder *kil-kil-kil* during display, but neither sound is particularly loud.

Occurrence in North America

Although there is no evidence so far of the Eurasian pochard breeding in North America, there is a long history of its irregular occurrences on Bering Sea islands (Pribilofs and St. Lawrence Island) and on islands in the western Aleutians.

Recent (to 2016) Bering Sea sight records reported via eBird include St. Paul Island (Pribilofs); reports have occurred during at least ten years since 1912: 1912, 1966, 1973, 1979, 1985, 1987, 1995, 1998, 2006, and 2007. St. George Island (Pribilofs) sight records include April 24, 1968, and May 22, 2016. The species has also been recently observed at least once on St. Lawrence Island (May 17, 2006).

Western Aleutian Islands eBird records from Adak Island include Sweeper Cove on June 1, 2016, and Small Boat Basin, various times from 1995 to 2012. Attu Island records include Murder Point on May 16, 1982. It was also reported at least three times from Casco Cove: May 16, 1989; June 12, 1991; and from May 25 to June 3, 1998.

There are a few mainland Alaska eBird records from Nome (June 5, 1994) or from the general vicinity of Nome—on Council Road at Mile 29.5 (June 4 and 5, 2014) and Safety Sound Bridge to Solomon (June 4 and 5, 2014) (a repeat of Nome vicinity sightings?).

Records from south of the Alaska Peninsula include Middleton Island (July 7 and September 9, 2012) and several sightings from lakes near the Kodiak Island airport (December 28, 2015, to March 22, 2016).

There had been at least three records from California as of 2016, the most recent on December 20, 2016 (American Birding Association blog, December 2016). Two males were reported from Orange County in 1994, and a male was seen in San Bernardino County in February 1989 (Patten, 1993). Pyscgock (2013) reported a Eurasian pochard from Lake Champlain, Vermont, in January 2013. It was the second of only two eastern North America records, the other from Quebec, during the spring of 2008. There is also a Canadian record from Saskatchewan.

Redhead
Aythya americana (Eyton) 1838

Other vernacular names. Red-headed duck, red-headed pochard

Range. Breeds from the Yukon-Kuskokwim Delta east locally in Alaska, the Northwest Territories and southeast through British Columbia and the prairie provinces to southern Ontario and southern Quebec to the St. Lawrence River valley. South of Canada it breeds from eastern Washington and southeastern Oregon to central California, and east through Nevada, northern Utah, northern Colorado, and the Nebraska Sandhills to northern Iowa and Minnesota. Some breeding also occurs farther east through the Great Lakes states, with local or occasional breeding elsewhere. Winters from the southern part of its breeding range from British Columbia east over the southern Great Plains to New England and south through the Atlantic states and Gulf coast west and south to coastal Tabasco and Yucatan.

Subspecies. None recognized.

Measurements. *Folded wing:* Delacour (1959): Males 230–242 mm, females 210–230 mm. Kear (2005): 210–256 mm, ave. of 2,131, 233 mm; females 206–243 mm, ave. of 1,255, 224 mm.
 Culmen: Delacour (1959): Males 45–50 mm, females 44–47 mm. Kear (2005): 41–62 mm, ave. of 2,094, 48 mm; females 41–62 mm, ave. of 1,253, 48 mm.

Weights (mass). Nelson and Martin (1953): 82 males, ave. 2.5 lb. (1,133 g), max. 3 lb. (1,361 g); 40 females, ave. 2.2 lb. (997 g), max. 2.9 lb. (1,314 g). Combined data of Bellrose and Hawkins (1947) and Jahn and Hunt (1964): 4 adult males (fall), ave. 2.39 lb. (1,084 g); 14 immature males, ave. 2.22 lb. (1,006 g); 6 adult females, ave. 2.28 lb. (1,034 g); 5 immature females, ave. 2.17 lb. (984 g).

Identification

In the hand. This duck is easily recognized as a pochard by its lobed hind toe and generally broad, flattened bill. *Males* in nuptial plumage might be identified by their uniformly coppery red head and yellow eyes, and by their flattened bluish bills with a pale subterminal band and a blackish tip. The black breast and the uniformly gray speculum, of nearly the same color as the upper wing-coverts, are similar to those of the canvasback, but the black breast extends from the wings to the foreneck, and the upper wing-coverts are slightly darker, rather than lighter, than the secondaries. *Females* can be separated from female canvasbacks by their shorter bills (under 50 mm) and more rounded head profile (see canvasback account) and from female ring-necked ducks by their longer wings (over 200 mm), black margined inner secondaries, less definite eye-ring and eye-stripe, and the usual presence of white flecking on their scapulars (see also the ring-necked duck account).

In the field. On the water, redheads appear to be shorter bodied and shorter necked than canvasbacks, and have a shorter and more rounded head profile. Males have a brighter, more coppery head color, and the backs and sides of the body are medium gray rather than whitish. During late winter and spring the male courtship call of redheads is frequent and audible for long distances; it is a unique catlike *meow* that few would attribute to a duck.

In flight, male redheads appear mostly grayish to white underneath, except for the black breast (which extends back to the leading edge of the wings) and brownish to coppery-red head. Their shorter neck and greater amount of black on the breast are the best means of distinction from male canvasbacks. Like canvasbacks, redheads fly with strong, rapid wingbeats, in a swift flight with relatively little dodging or flaring, but they are somewhat more agile in flight than are canvasbacks. Females are more uniformly brownish on the head, breast, sides, and back, lacking the more two-toned effect of female canvasbacks. Females likewise appear white on the abdomen and the underwing surface, and the brown color of the head and breast extends back in an unbroken manner along the flanks. Like most pochards, females rarely utter any loud calls that could be useful for field identification.

Age and Sex Criteria

Sex determination. In any adult plumage, a pale, yellowish eye indicates a male, as do vermiculations anywhere except on the scapulars, where females are sometimes slightly vermiculated. However, only males are vermiculated near the tips of the tertials (Carney, 1964).

Age determination. The greater secondary and tertial coverts of adults are broad and rounded. In males they are also heavily flecked with white, whereas in females they are unflecked or only faintly flecked near their edges. Juveniles are very female-like, but their greater coverts are narrower, squared-off, and often somewhat frayed with pale edges. Juvenal tertials, until molted, indicate immaturity by their frayed, pointed tips and brownish gray coloration (Carney, 1964). The blunt-tipped juvenal tail feathers are dropped in no apparent sequence at 3.5 to 7 months of age (Weller, 1957). Weller also reported that young males can be recognized by their reduced area of black in the breast region as compared with older birds, and young females usually exhibit speckled buffy brown on their under tail-coverts, whereas older females have brownish olive patches there.

Distribution and Habitat

Breeding distribution and habitat. In Alaska breeding redheads are very local but are known to nest in the Tetlin and Minto areas, along the Tanana River, and in the Fort Yukon area of the Yukon and Porcupine Rivers.

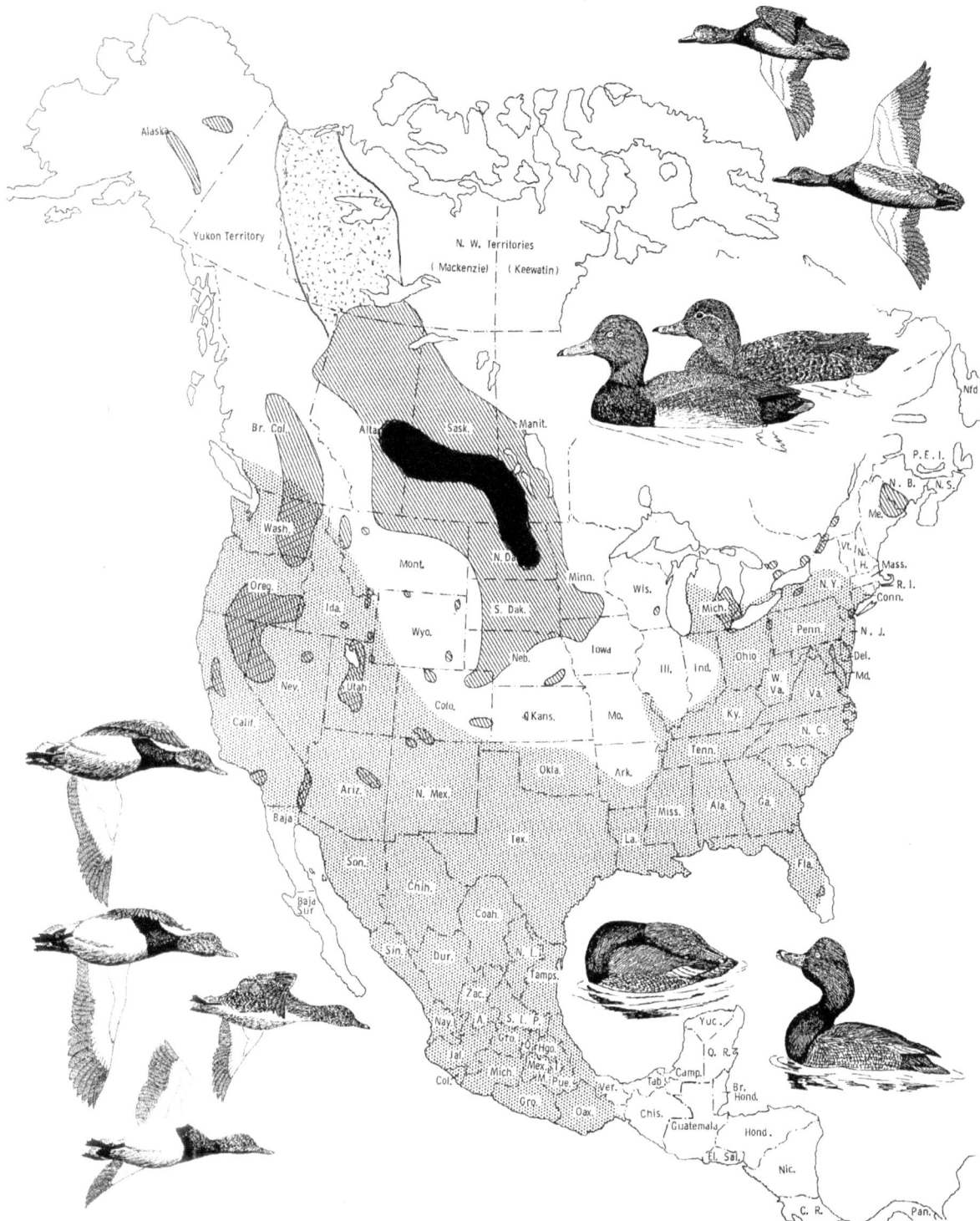

The breeding (hatched, with denser concentrations inked), wintering (shaded), and marginal (stippled) range of the redhead.

In Canada redheads breed in the intermontane region of British Columbia and are particularly prevalent in the prairie provinces of Alberta, Saskatchewan, and Manitoba, extending locally northward as far as Great Slave Lake, Northwest Territories. There are several small breeding localities in the southern part of Ontario, including Lake St. Clair, Charter Island, Luther Marsh, and Toronto Island (Godfrey, 1986). In Quebec redheads have bred at Lake St. Francis and perhaps also on the St. Lawrence River near Trois-Rivières; the latter might be the result of releasing captive birds (Weller, 1964). Breeding has also been recorded in New Brunswick, which evidently is the eastern limit of the species' breeding range.

South of Canada the breeding range of the redhead is discontinuous and declining but is centered in the prairie potholes region of the Dakotas and Canada's prairie provinces. Small local breeding populations occur in nearly all of the western states from Washington south to California and east as far as Kansas, Nebraska, Iowa, and Minnesota. The southernmost US breeding might be in southeastern New Mexico, but some breeding also occurs in the playa lakes of western Texas and in the transvolcanic belt of central Mexico.

In Minnesota the species reaches the eastern limit of its major breeding range and is about the fifth most common breeding duck (Lee et al., 1964a), with a population of about 9,000 birds (Cordts, 2010). In Iowa it breeds in a few northwestern and north-central counties (Weller, 1964; Dinsmore et al., 1984), and in Wisconsin it is a localized breeder (Jahn et al., 1964). In Michigan it breeds on bays and marshes of the Great Lakes, and on some shallow inland lakes (Brewer, McKeek, and Adams, 1991). There are older breeding records for Ohio and Pennsylvania, and in New York there have been breeding records at Jamaica Bay, Long Island, as well as in eastern New York (McGowan and Corwin, 2008).

Weller (1964) described the redhead's breeding habitat as nonforested country with water areas sufficiently deep to provide permanent, fairly dense emergent vegetation for nesting cover. Weller believed that this species evolved in the alkaline water areas of the American Southwest, and it still attains its highest breeding densities in alkaline wetlands.

In Minnesota redheads usually nest in wet emergent vegetation from 20 to 40 inches tall, typically among cattails or similarly high vegetation around deep potholes that have some open water present (Lee et al., 1964a). Lokemoen (1966) found that redheads preferred to nest in potholes at least one acre in area, and that potholes most suitable for brood rearing were at least this large and were deeper than those used for nesting.

Low (1945) observed that the highest nesting densities in Iowa occurred where about 10 to 25 percent of the habitat consisted of open water; the areas of open water used for landing and taking off were at least a square rod (about 5 square meters) in size, and usually 3 to 4 rods (about 50–65 feet) square. Water depth in nesting areas appeared to be more important than the presence of specific plant species, with a water depth of about 9 inches at the nest site seemingly favored. Water areas used for brood rearing were larger and deeper, and more had open water than those used for nesting.

Population. Although major declines in the redhead population occurred during the first third of the twentieth century, including the drought years of the 1930s, it was not until the mid-1970s that the species had begun to make a slow recovery. Spring breeding surveys of the "Traditional" survey route of western and central Canada and the northern prairie states indicated average populations of about 550,000 birds in the

Fig. 13. Redhead, adult male swimming

1955–69 period, 640,000 in the 1970s, 590,000 in the 1980s, 728,000 in the 1990s, 670,000 in 2000–05, and about 1,000,000 from 2006 to 2010 (Baldassarre, 2014). The 2014 estimate was 1.3 million, or 85 percent above its long-term average (Zimpfer et al., 2014), and the 2016 population was also estimated at 1,300,000 birds (US Fish and Wildlife Service, 2016).

Recent hunter-kill estimates in the United States averaged about 140,000 birds from 1999 to 2008, and about 17,500 in Canada for the same period. In the United States almost half of the total kill occurred in the Central Flyway, about a third in the Mississippi Flyway, 13 percent in the Pacific Flyway, and 4 percent in the Atlantic Flyway. Nearly half of the Canadian kill occurred in Manitoba (Baldassarre, 2014).

Wintering distribution and habitat. Weller has provided an excellent summary of the distribution and relative abundance of redheads in their major North American wintering areas. He determined that 78 percent of the wintering birds, based on 1951–56 winter inventory surveys, were concentrated along the Laguna Madre of coastal Texas and adjacent Tamaulipas. Another 11.9 percent occurred from the Chesapeake

Bay region south to Pamlico Sound, and coastal Florida supported about 5 percent. The remaining 5 percent occurred on the western coast of Mexico, in California, along the southern Great Lakes, and other minor wintering areas.

Midwinter surveys of redheads from 2000 to 2010 in the United States averaged about 390,000 birds, with 62 percent in the Central Flyway (nearly all in Texas), 23 percent in the Atlantic Flyway, 8 percent in the Pacific Flyway, and 6.5 percent in the Mississippi Flyway. The Laguna Madre region of Texas and adjoining Tamaulipas averaged about 240,000 birds in the Texas portion and 235,000 in the Mexican component (Baldassarre, 2014).

Weller characterized typical wintering areas as large bodies of water along the coast that are well protected from heavy wave action. They are often fairly shallow, and they might be brackish or highly saline, as in the case of the Laguna Madre. Stewart (1962) indicated that in the Chesapeake Bay area redheads are most numerous during winter in brackish estuarine bays containing extensive beds of clasping-leaf and sago pondweeds (*Potamogeton perfoliatus* and *P. pectinatus*), wigeon grass (*Ruppia*), and eelgrass (*Zostera*).

During spring and fall migration, redheads evidently prefer fresh and slightly brackish estuarine bays, and concentrate in areas having an abundance of submerged plants such as wild celery (*Vallisneria*) and naiad (*Najas*). They also use more brackish areas like those typical of wintering birds, but concentrate on freshwater areas. Stewart suggested that seasonal shifts of habitat might be related to weather severity and resulting ice conditions in different areas during winter.

General Biology

Age at maturity. Ferguson (1966) noted that only 6 of 19 aviculturists reported breeding by captive redheads in their first year of life, but in part this apparent delayed maturity might reflect the general difficulties of breeding this species under captive conditions. Since Weller (1965) noted that all the wild females he observed had established pair-bonds by the time of their arrival at breeding areas, it seems probable that many of them at least attempt to nest during their first year. Possibly the yearling birds are responsible for much of the parasitic egg-laying found in this species, as a result of incompletely matured nest building and brooding tendencies.

Pair-bond pattern. Pair-bonds are established yearly, after a rather prolonged period of social courtship (Weller, 1965, 1967). Pair formation begins as early as late December or January, and normally persists until about the beginning of incubation (Oring, 1964), although Hochbaum (1944) recorded a single case of the pair-bond apparently persisting until after hatching had occurred.

Nest location. Nests are typically found over standing water in emergent vegetation or on a mass of plant material surrounded by water. In Minnesota wet cattail stands are the most common nest sites of redheads, although other emergent species are also used (Lee et al., 1964a, 1964b). The average height of vegetation above the water surface in a sample of Minnesota nests was 29 inches, with a range of 20 to 40 inches. This average

Redhead, breeding pair swimming

height was slightly less than that of canvasback nests and more than that of ring-necked ducks. Nine redhead nests averaged 9.7 yards from open water, with almost half within 5 yards of open water and none beyond 50 yards. Canvasback and ring-necked duck nests were very similar to those of redheads in this regard.

Lokemoen (1966) analyzed nesting preferences of redheads in Montana and found that hardstem bulrush (*Scirpus acutus*) was the most highly preferred cover, but that because of its greater abundance cattail was most commonly used. Baltic rush (*Juncus balticus*) and spike rush (*Eleocharis*) were third and fourth place in the preference scale. Large stands and wide bands of emergent vegetation were preferred over smaller or more disrupted stands for nesting, and water depth at the nest site averaged ten inches. Potholes larger than one acre in size were preferred over smaller ones for nesting sites, and none under one-fourth acre in size were utilized. Williams and Marshall (1938) also found hardstem bulrush to be the most highly preferred nesting cover, with alkali bulrush (*S. paludosus*) scarcely utilized and both cattail and phragmites having only limited usage. Miller and Collins (1954) likewise determined that hardstem bulrush from 2 to 10 feet high is preferred nesting cover.

Clutch size. Weller (1959) calculated that a total of 1,380 redhead nests reported in eight different studies had an overall average clutch size of 10.8 eggs, with averages of individual studies ranging from 8.9 to 13.8 eggs. However, Weller found that of 17 nests with clutches laid by a single hen, none exceeded 9 eggs, and the average clutch size was only slightly more than 7 eggs. Lokemoen (1966) estimated an average clutch of 7.9 eggs for nonparasitized nests and reported finding 23 probable renesting attempts. Weller considered that renesting was unlikely to be important in redheads owing to the lateness of the peak in initial nesting attempts.

Incubation period. The redhead incubation period is reported as 24 days by Hochbaum (1944). Weller (1957) reported it to be 24 to 28 days, and Low (1945) stated that five nests he studied had an average incubation period of 24 days, although one other nest required 28 days.

Fledging period. Weller (1957) determined that hand-reared birds fledged at 56 to 73 days of age. Lightbody (1985) also reported fledging by hand-raised birds at 63 days. Low reported flying by wild birds at 70 to 84 days.

Nest and egg losses. Weller (1964) estimated an average nesting success of 53 percent for 503 nests found among six different field studies. He also (1959) calculated an average hatching success of 32 percent for a total of 10,802 eggs that were monitored in six studies. He believed that only 50 to 60 percent of the female population built nests, and he found that eggs laid by parasitic females had a low hatching success. More recently, Lokemoen (1966) reported 15.2 percent nesting success and 9.9 percent hatching success for the eggs in 138 nests, with desertion and communal nesting attempts accounting for more than half of the failures. Mammals (mostly striped skunks) and birds (black-billed magpies and American crows) also accounted for some nest losses.

In total, Weller (1964) believed that the estimated 60 percent of the female redheads in his study population that attempted to nest hatched an average of 3.4 young per nest, and that about only one egg laid by each parasitic female hatched, assuming a 10 to 15 percent hatching success of such eggs.

Juvenile mortality. Prefledging mortality of ducklings is still not well known, but Low (1945) estimated there might be a 30 percent loss of young during the first six weeks of life. Weller (1964) provided brood size data for well-grown broods that suggest an even higher survival rate, but brood mergers very probably reduce the reliability of such data.

First-year mortality of redheads is extremely high and might average about 75 percent for the year following banding (Hickey, 1952). Rienecker (1968) calculated an even higher mortality rate (78.7 percent) for first-year birds, as Brakhage (1953) did for wild-trapped (80 percent) and hand-reared (94 percent) birds. Females of both the immature and mature age classes are considerably more vulnerable than males to gunning mortality (Benson and DeGraff, 1968) and additionally are more greatly exposed to dangers of predation during nesting.

Redhead, breeding pair swimming

Adult mortality. Adult annual mortality rates of redheads have been estimated by Hickey (1952) at about 55 percent and by Rienecker (1968) at 41 percent. Longwell and Stotts (1959) estimated a 44 percent mortality for Chesapeake Bay redheads. Lee et al. (1964b) estimated a 62 percent adult mortality, as compared with an estimated 80 percent rate for first-year birds. These figures, although not in extremely close agreement, all suggest a dangerously high rate for adults as well. In contrast to Rienecker's conclusion, Geis and Crissey (1969) judged that highly restrictive hunting regulations resulted in significant reductions in the mortality rates of redheads and canvasbacks. The maximum known record of survival in the wild is 22 years, 7 months.

General Ecology

Food and foraging. The summaries by Martin et al. (1951) and Cottam (1939) of redhead foods indicate that the vegetative parts and seeds of pondweeds (*Potamogeton*), wild rice (*Zizania*), wild celery (*Vallisneria*), and wigeon grass (*Ruppia*); the seeds of bulrushes (*Scirpus*); and the vegetative parts of muskgrass (*Chara*) are major foods in various parts of the country. In the important wintering area Laguna Madre, McMahan (1970) found that more than 90 percent of the volume of food materials in 104 redhead samples consisted of wigeon grass and shoalgrass (*Diplantera*), with the latter occurring in 83 percent of the samples and alone constituting 84.2 percent of the food volume. Small gastropod and pelecypod mollusks made up the relatively insignificant proportion of animal materials that were found. Lynch (1968) noted the importance of shoalgrass to wintering redheads throughout the Gulf coast.

Stewart (1962) summarized the foods of redheads from the Chesapeake Bay region, based on a sample of 99 birds. There, the leaves, stems, rootstalks, and seeds of submerged plants were also the principal foods, but the food species differed considerably. In freshwater estuaries various pondweeds and naiad (*Najas*) were major foods; in brackish estuaries eelgrass (*Zostera*) and clasping-leaf pondweed (*P. perfoliatus*) were most important; and in samples from saltwater estuaries these two species plus wigeon grass had been taken, as well as bait corn and sorghum.

The findings of Bartonek and Hickey (1969) on summer-collected redheads on their breeding grounds in Manitoba indicate a higher usage of animal materials by both adult and young birds than had been generally appreciated. Aquatic invertebrates form the bulk of spring and summer foods, especially cladocerans, gastropod mollusks, and the larvae of Trichoptera (caddisflies) and Tendipedidae (midges).

Sociality, densities, territoriality. To a degree that seems stronger than in the canvasback, the redhead appears to exhibit a sociality on the breeding grounds that might in part be related to its semiparasitic nesting tendencies. These tendencies might partly result from the redhead's specialized requirements for nesting sites, which cause a concentration of nests in the limited suitable habitat. Williams and Marshall (1938) estimated an average nesting density of 0.11 redhead nests per acre in 3,000 acres of total nesting cover, but up to 11 nests per acre in a two-acre area of alkali and hardstem bulrushes, their preferred nesting cover. Vermeer (1970) reported an average redhead nesting density of 0.11 nests per acre and noted redheads to be among the species of ducks he found to nest in higher densities among tern colonies than in areas where they were absent.

Densities over larger areas of breeding habitat are much lower than those just mentioned. Stoudt (1969) reported that in five prairie study areas in Canada and South Dakota, the peak density of redheads varied from 1 to 6 pairs per square mile. Lokemoen (1966) reported an unusually high density of 25 pairs per square mile on a 2,600-acre study area in western Montana. However, a high incidence of attempted communal nesting and nest desertion were associated with this remarkably high breeding density.

There is no evidence that redheads defend a nesting territory or their larger and more inclusive home range; Lokemoen (1966) noted that males did not defend any part of their home range. Hochbaum (1944) believed that redheads were the most tolerant of the breeding diving ducks in the Delta, Manitoba, region, based on the often close special associations of different pairs, with as many as three pairs simultaneously occupying a half-acre slough.

Redhead, breeding pair in flight

Interspecific relationships. The significant role that social parasitism of redheads plays in the breeding biology of other marsh-nesting species has been documented by Weller (1959), who noted that eight other species of ducks, as well as bitterns and coots, have been parasitized by redheads, and both Weller's and Erickson's (1948) studies indicated that social parasitism by redheads reduced the hatching success of canvasback eggs. Erickson also found a reduced nesting success for parasitized canvasback nests, as compared with nonparasitized nests.

Weller (1959) also noted that a number of other species of ducks, including the ruddy duck, mallard, lesser scaup, canvasback, fulvous whistling-duck, pintail, cinnamon teal, shoveler, and gadwall, have occasionally been reported to drop their eggs in redhead nests.

Redheads have the usual array of egg and duckling predators, although the fact that they normally nest over water and well away from shoreline probably reduces their losses to strictly terrestrial scavengers and predators. Keith (1961) reported that half the redhead nests he found in southeastern Alberta were on land,

and many of these were very poorly concealed. He noted that striped skunks destroyed a number of red-head nests, and Lokemoen (1966) also found that skunks were the major mammalian predators of redhead nests in Montana. Low (1945) noted that minks and American crows were responsible for nest losses in Iowa, whereas crows and black-billed magpies were noted by Lokemoen (1966) to be avian egg predators.

General activity patterns and movements. Home range estimates for redheads on their breeding grounds are still generally not available. Lokemoen (1966) stated that pairs moved an average of 180 yards (the variation among 11 pairs was 50–670 yards) from their "breeding-pair potholes" to nesting potholes.

Long-distance movements of redheads have been analyzed by Weller (1964). He documented the occurrence of a postseason adult molt migration in a northerly and somewhat easterly direction, as well as similar movements by juvenile birds. He also established the directions and relative magnitudes of spring and fall migratory movements, pointing out that the flyway concept is relatively meaningless in interpreting this species' movements. In contrast to the canvasback, which predominantly moves to the Atlantic coast or the Pacific coast for wintering, the vast majority of redheads undertake the relatively long flight over dry country to the Gulf coast. Weller attributes this difference in part to the hypothesized differences in areas of evolutionary origin of these two species.

Social and Sexual Behavior

Flocking behavior. Like canvasbacks, redheads often gather in fairly large flocks on lakes that provide protection and food, forming large "rafts" that might number in the hundreds or even thousands. During the winter and spring migration periods, these large groupings tend to fragment as pair-bonds are formed, and the unpaired birds congregate in courting party units. Low (1945) noted that spring migrant flocks usually did not exceed 25 individuals, and Weller (1967) mentioned that sometimes as many as 14 males were seen following a single unmated female. Shortly after arrival at the breeding grounds, the paired birds separate and disperse, and flocking behavior ceases until after the breeding season.

Pair-forming behavior. The pair-forming behavior of redheads is similar to that of canvasbacks and other pochard species (Johnsgard, 1965; Weller, 1967) and differs in quantitative rather than qualitative characteristics. The commonest male courtship call is a catlike note, uttered during neck-kinking (Fig. 14C–D) or a head-throw display (Fig. 14A–B). A softer call resembling coughing is also uttered, and aggressive neck-stretching by both sexes is frequent. Females perform inciting calls with alternate lateral and chin-lifting movements of the head, and a frequent male response to such inciting is to swim ahead and turn-the-back-of-the-head toward the inciting female (Fig. 14E). Weller (1967) noted that males on wintering areas were observed to "lead" females, and the latter's action in following them seemed to indicate a willingness to pair. This same combination of leading and following has been observed in captive birds (Johnsgard, 1965) and seems to represent a significant aspect of pair formation among both dabbling ducks and pochards.

Aerial chases, involving tail-pulling, are characteristic of birds on the breeding grounds but are rare during

Fig. 14. Sexual behavior of redhead (A–F) and ring-necked duck (G–H), including (A) head-throw, (B–D) kinked-neck call, (E) male leading inciting female, (F) postcopulatory bill-down posture, (G) head-throw, and (H) triangular-crest posture (after Johnsgard, 1965).

migration, suggesting that they do not play a role in the pair-formation process, which is virtually completed by the time of the birds' arrival at their nesting grounds. More probably, they are associated with chases of the female by strange drakes and represent attempted rapes.

Copulatory behavior. Copulation is normally preceded by alternate bill-dipping and dorsal-preening behavior on the part of the male or, at times, by both male and female. The female then assumes a receptive posture and is immediately mounted by the male. Following treading, the male normally utters a single note as he releases his grip on the female's nape, and he swims away in a stereotyped bill-down posture (Fig. 14F). The female sometimes assumes this same posture for a short time before she begins to bathe (Johnsgard, 1965).

Nesting and brooding behavior. Low (1945) analyzed the nesting and incubating behavior of redheads. He found that nest building began two days to a week before egg laying. Eggs were deposited in the nest at any time of the day, as Weller (1959) later confirmed, although most eggs are apparently laid before noon. One to two additional days are required to lay a clutch than the number of eggs present, indicating an egg-laying rate of slightly more than one day per egg. Incubation might begin as late as 24 to 48 hours after the last egg is laid. During incubation the females that Low studied left the nest an average of six times a day. Renesting females not only left their nests more often but also spent less total time on the nest than those making initial nesting attempts.

Pipping requires 16 to 18 hours, and Weller (1959) noted that during this period the female begins to utter low notes that probably serve to "imprint" the ducklings on their mother. Weller watched one brood that left its nest when the young were no more than 47 hours old. Redhead females are well known to be relatively poor parents, rarely feigning injury when a family is approached and often deserting their brood while it is still fairly young. Low (1945) said that the young were usually abandoned by the time they were 7 to 8 weeks old, before they were able to fly.

Postbreeding behavior. Male redheads usually abandon their females fairly early in the incubation period and soon begin to gather in groups prior to their postnuptial molt. At least in some areas, a fairly long molt migration might be undertaken by such birds to more northerly areas to certain large, shallow lakes such as Lake Winnipegosis, Manitoba (Weller, 1964). Young redheads might also move considerably during late summer and autumn following fledging, and also often range far to the north of the place where they were reared. There is no strong evidence favoring a major differential migration of the sexes during fall, but the greater vulnerability of females to gunning results in a high proportion of this sex being shot during fall migration. Rienecker (1968) noted, however, that males range farther than females in their migratory movements.

Ring-necked Duck
Aythya collaris (Donovan) 1809

Other vernacular names. Blackjack, ring-billed duck, ring-bill duck

Range. Breeds from central Alaska and the Northwest Territories southeast through the forested regions of central and southern Canada east to Nova Scotia and western Labrador, south locally to northern California, Idaho, western Wyoming, western Colorado, northeastern South Dakota, Iowa (rare), the Great Lakes states, New York, and New England. Winters along the Pacific coast from British Columbia to Baja, California, central Mexico and adjoining Central America, east through the Gulf States and Florida, and north along the Atlantic coast to Massachusetts.

Subspecies. None recognized.

Measurements. *Folded wing:* Delacour (1959): Males 195–206 mm; females 185–195 mm. Kear (2005): Males 191–220 mm, ave. of 200, 205 mm; females 178–210 mm, ave. of 196, 196 mm.

Culmen: Delacour (1959): Males 45–50 mm; females 43–46 mm. Kear (2005): Males 45–52.2 mm, ave. of 264, 49.1 mm; females 44.5–50.2 mm, ave. of 158, 47.1 mm.

Weights (mass). Nelson and Martin (1953): 285 males, ave. 1.6 lb. (725 g), max. 2.4 lb. (1,087 g); 151 females, ave. 1.5 lb. (679 g), max. 2.6 lb. (1,178 g). Combined data of Bellrose and Hawkins (1947) and Jahn and Hunt (1964): 17 adult males (fall), ave. 1.74 lb. (789 g); 33 immature males, ave. 1.53 lb. (694 g); 15 adult females, ave. 1.51 lb. (685 g); 29 immature females ave. 1.49 lb. (676 g).

Identification

In the hand. Ring-necked ducks are often misidentified by hunters, the males usually being confused with scaup, and the females with scaup or redheads. The pale whitish ring near the tip of the bill separates both sexes from scaup, as does the absence of predominantly white secondary feathers. The male ring-necked duck might be readily distinguished from redheads or canvasbacks by its darker, rather glossy, greenish black upper wing-coverts and tertials, which lack any light gray and black vermiculations.

Females are much more difficult to separate, for although ring-necks lack the long, sloping bill of female canvasbacks, redheads also have a whitish band near the tip of the bill. Nevertheless, unlike female redheads, female ring-necked ducks have secondaries that are more distinctly grayish than are the relatively brown coverts, and a white eye-stripe and eye-ring are more evident. The wings are shorter (folded wing less than 200 mm vs. at least 210 mm in female redheads), and the scapulars are never flecked or vermiculated with whitish.

In the field. When in nuptial plumage, the male ring-necked duck on the water is the only North American diving duck that has a black back and breast pattern, with a vertical white bar extending upward in front of the folded wing. The rare tufted duck also has a black back and breast but lacks the white bar and has a much longer and thinner crest than does the ring-necked duck. Its white ring near the tip of the bill is often apparent at close range, but the chestnut ring at the base of the neck is rarely visible.

Females on the water are probably best identified by their association with males, but usually they exhibit a white eye-ring and posterior eye-stripe as well as a white ring near the tip of the bill. Females lack the white forehead of scaups but do have distinctly pale areas near the base of the bill. In flight, ring-necked ducks resemble scaups but lack white wing-stripes, and their darker back and upper-wing coloration also serves to separate them from redheads and canvasback. Ring-necked ducks are relatively quiet, and the courting calls of the male include a soft breathing note and a louder whistling sound that is difficult to characterize, both of which are uttered only on the water.

Age and Sex Criteria

Sex determination. Males have yellowish rather than brownish eyes, and a pale area at the base of the bill. Vermiculated flanks or black feathers on the head, breast, or back also indicate a male. Sex determination by wing characters alone is difficult, but the tertials of males are more shiny greenish black and more pointed than those of females, and the secondary coverts are darker and might be slightly glossy.

Age determination. Juveniles are female-like but have less grayish cheeks, a more spotted belly, and darker irises (Baldassarre, 2014). The tail feathers of juveniles have notched tips. Juvenal tertials are pointed, straight, and usually badly frayed, whereas those of adults are more rounded and usually slightly curved. Likewise the greater and middle coverts of juveniles are relatively narrow, frayed, and rough (Carney, 1964).

Distribution and Habitat

Breeding distribution and habitat. Mendall (1958) documented the breeding range of this strictly North American species; his important work might be consulted for details of historic distribution.

The ring-necked duck did not until recently breed in Alaska, but it has been observed from the Bering Sea to the Canadian border (Hansen, 1960). The species has been increasing in Alaska for many decades, and by the 1960s several breeding records had accrued (White and Haugh, 1969). In Canada it is for the most part restricted to the area south of latitude 60°N, with its northernmost limits near Fort Simpson and lower Slave River (Godfrey, 1966). Otherwise, it breeds in the Cariboo Parklands of British Columbia; over much of Alberta, Saskatchewan, and Manitoba north of the "prairie pothole" country; in Ontario from Hudson Bay south to the Great Lakes; in southern Quebec; in the Maritime Provinces; and in southern Newfoundland.

To the south of Canada, the species is primarily found in the Great Lakes and New England regions, but isolated breeding does occur elsewhere. Breeding occurs in northeastern Washington as well as along the

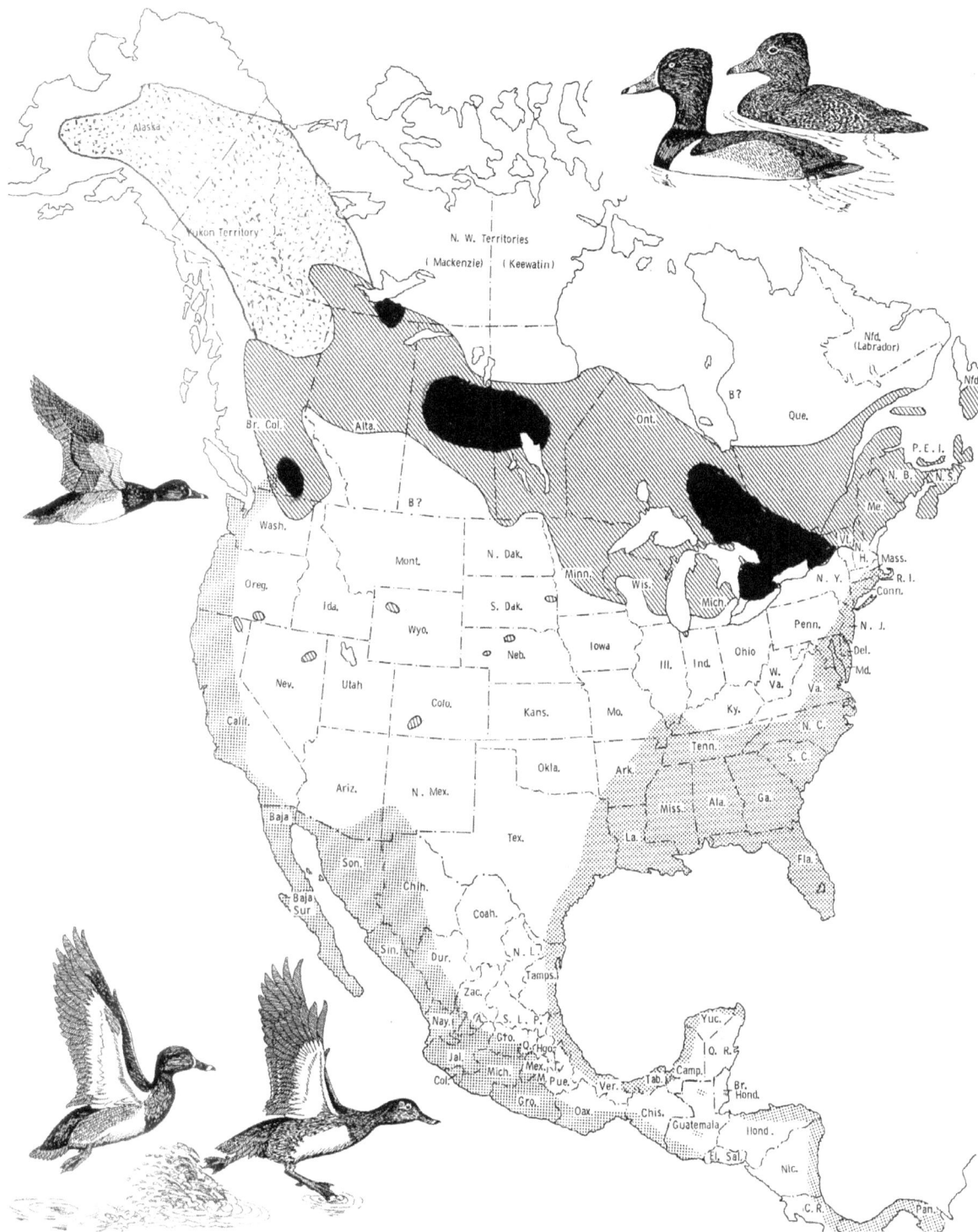

The breeding (hatched, with denser concentrations inked), wintering (shaded), and marginal (stippled) range of the ring-necked duck.

Puget Sound coast (Hohman and Eberhardt, 1998). In Oregon there is local breeding in the Cascade Mountains, with some breeding records for upper Klamath Lake and Malheur National Wildlife Refuge (Gilligan et al., 1994). Local breeding has also been documented for Nevada, Montana, eastern Idaho, western Wyoming, western Colorado, and Arizona. Ring-necked ducks were once thought to breed locally in Nebraska's Sandhills lakes (Rapp et al., 1950), but no proof has ever been found. Breeding in South Dakota is rare; in North Dakota they breed locally in the eastern and northeastern portions of the state (Stewart, 1968). In northwestern and central Iowa they are also local nesters (Dinsmore et al., 1968).

In Minnesota and Wisconsin the ring-necked duck has long ranked third in number, behind the blue-winged teal and the mallard, among the breeding birds of these states (Lee et al., 1964a; Jahn and Hunt, 1964), and in Michigan it is a common breeder in the northern two-thirds of the state (Brewer, McKeek, and Adams 1991). It has also been recorded breeding in Illinois, Indiana, and Pennsylvania, and in New York it breeds over a 500-square-mile area of the Adirondacks (Foley, 1960). Vermont, New Hampshire, and Maine probably represent its southern limit of regular breeding in New England, but it has bred a few times in Massachusetts (Mendall, 1958).

Mendall has characterized the favored breeding habitat as sedge-meadow marshes and bogs, ranging in size from an acre to nearly 2,000 acres. Shallow freshwater marshes, swamps, and bogs are all used by ring-necks, and bogs are especially favored, particularly those with sweet gale (*Myrica*) or leatherleaf (*Chamaedaphne*) cover. Further, white water lily (*Nymphaea odorata*) and water shield (*Brasenia schreberi*) are frequent associate plants of nesting birds in Maine, as are yellow water lilies (*Nuphar*) in Washington. Fresh water or acidic areas are apparently preferred over brackish or saline waters; a pH range of 5.5 to 6.8 is typical of breeding habitats.

Population. The ring-necked duck population has generally fared better than either of the two scaups, the canvasback, or redhead. The long-term average counts for the "Traditional" survey route (1955–2010) and the eastern survey (1990–2011) of the US Fish and Wildlife Service and the Canadian Wildlife Service collectively indicate a population of about 1.2 million birds. The long-term population trend has also been upward, especially for the traditional Alaska to western Canada route (Baldassarre, 2014; Canadian Wildlife Service Waterfowl Committee, 2013). The US hunter kill during the 2014 and 2015 seasons averaged about 498,000 birds (Raftovich, Chandler, and Wilkins, 2015), the largest total among all of the pochards and sea ducks, and similar to the average of about 494,500 for the 1999–2008 period.

Wintering distribution and habitat. In Canada, ring-necked ducks regularly winter in southwestern British Columbia, occasionally occur in southern Ontario, and rarely winter in Nova Scotia (Godfrey, 1985).

In recent (2000–10) midwinter surveys, nearly 50 percent of the wintering ring-necked duck population has been seen in the Mississippi Flyway (mostly in Louisiana), almost 25 percent in the Central Flyway (nearly all in Texas), and 16 percent in the Atlantic Flyway, and the remainders were in the western states. In Mexico they are mostly limited to the Gulf coast region, with major concentrations from Tamaulipas to northern Yucatan, and especially in the Laguna de Alvarado, Veracruz (Leopold, 1959; Baldassarre, 2014).

Fig. 15. Ring-necked duck, adult male swimming

They also winter along the Caribbean lowlands through Honduras at least as far as Panama, although in small numbers.

The Gulf coast of Texas supports many wintering ring-necks, but far fewer than does the corresponding area of Louisiana, which, with Tennessee, has the largest numbers of wintering birds in the Mississippi Flyway. In the Atlantic Flyway, the species is widely dispersed during winter on marshes, lakes, ponds, and reservoirs throughout the south, but peak concentrations probably occur in South Carolina, Georgia, Florida, and Alabama (Mendall, 1958; Addy, 1964). Some birds winter as far north as Chesapeake Bay, and very limited numbers occur locally even farther north.

In the Chesapeake Bay area, the preferred habitats of migrant and wintering ring-necks consist of fresh or slightly brackish estuarine bays and interior impoundments, with movement to moderately brackish waters during colder periods (Stewart, 1962). Mendall observed that on wintering areas the birds remain partial to shallow, acid marshes. They do also use coastal lagoons, where they often associate with scaup, but they generally select less brackish conditions than do scaup.

General Biology

Age at maturity. Ferguson (1966) noted that only one of 11 aviculturists observed breeding by captive ring-necked ducks in their first year of life, but nevertheless it is generally assumed that wild birds attain sexual maturity within a year of hatching, and that most breed at that age.

Pair-bond pattern. Pair-bonds evidently begin to be formed on wintering grounds, probably starting in January or February (Weller, 1965). The pair-bond typically breaks during the last week of incubation, or at the latest very shortly after hatching (Mendall, 1958). The association of apparently paired birds during fall migration might indicate the re-forming of bonds of previously paired birds, but this point has not been established.

Nest location. Mendall (1958) noted that of 518 nest sites monitored almost half were on floating islands, nearly 40 percent were on hummocks or clumps in open marsh, 9 percent were on solid islands, and the remaining few were on floating logs, in woodland swale, or in dry meadow. Only a single nest was on a dry site, and only two were in emergent vegetation, but the distance to open, permanent water averaged only 27 yards and ranged up to 400 yards. About 70 percent were within 15 yards of water sufficiently open for birds to land and take off. Perhaps the most important site criterion was the presence of a reasonably dry site with suitable cover fairly close to water of swimming depth. About 70 percent of the nests were in a mixture of sedge (*Carex*), sweet gale, and leatherleaf vegetation, and another 10 percent were in a mixture of sedges and other plants. More nests were found under sedge than under any other growth of a single plant species, but most nests were placed in mixed cover types. There was no evident relationship to distance from shoreline or woods, but small clumps of nesting cover seemed to support more nests than did larger ones.

Clutch size. Mendall (1958) calculated that the average size of 423 completed clutches was 9.0 eggs, with an observed range of 6 to 14. Renest clutches averaged about two eggs fewer (6.96), with nearly half of the observed cases having 7 eggs present. Eggs are apparently laid at the rate of one per day. Hunt and Anderson (1965) noted a reduction in average clutch size from 7.9 eggs in eight initial nests, to 7.8 eggs in eight second nests, and 7 eggs in one third nesting. Hunt and Anderson found that eight of ten marked females attempted to renest following nest loss, and one attempted a second renest.

Incubation period. Observed incubation periods on naturally incubated eggs have ranged from 25 to 29 days, with most clutches hatching after 26 to 27 days (Mendall, 1959).

Fledging period. Mendall reported a fledging period of 49 to 56 days for wild ring-necked ducks, which is a surprisingly short fledging period for any pochard species.

Nest and egg losses. Mendall estimated that 70 percent of 485 first nests under study hatched, whereas 61 percent of 52 renestings hatched. This relatively high nest success was associated with a very low nest

Fig. 16. Ring-necked duck, pair swimming

desertion rate, and most of the losses were attributed to predation. Major mammalian egg predators were minks, raccoons, and foxes, whereas crows, ravens, and northern harriers were the primary avian predators. Minks, American crows, common ravens, and raccoons also accounted for over 70 percent of the predation losses and probably contributed to the 19 percent loss by unknown predators.

Juvenile mortality. Mendall found that the average brood size at hatching was 8.4 young, whereas the average of well-feathered (class III) broods was 5.2 young. Evidently the highest brood losses occur in the first 48 hours of life, and thereafter the mortality is fairly low. Some losses were definitely attributed to snapping turtles, and this species was believed responsible for considerable brood mortality in some areas. Jahn and Hunt (1964) summarized data from a variety of studies that indicated an average of about 6 ducklings per brood surviving to near the flight state. They also judged that about half the females succeeded in producing broods.

Mortality rates of birds banded as juveniles and recovered in their first year after banding are apparently high. Lee et al. (1964b) calculated a 75.7 percent mortality rate for such birds, and Jahn and Hunt (1964) estimated a similar 70 percent annual immature mortality rate.

Adult mortality. Lee et al. (1964b) calculated a 66 percent annual mortality rate for ring-necked ducks recovered from one to five years after banding, while Jahn and Hunt (1964) estimated a 50 percent annual adult mortality rate. Conroy and Eberhardt (1983) analyzed survival rates of nearly 32,000 birds banded in eastern North America. They determined the highest annual survival rates for males (62.9 to 69.2 percent), followed by 48.0 to 58.4 percent for females, and 36.3 percent for immatures of both sexes.

General Ecology

Food and foraging. Martin et al. (1951) and Cottam (1939) both reported that the seeds of water shield, the seeds and vegetative parts of pondweeds, and the seeds or vegetative parts of various other submerged or emergent aquatic plants are consumed by ring-necked ducks in considerable quantities. Additionally, animal materials such as insects, mollusks, and other aquatic animal life are taken in substantial amounts, averaging about a quarter of the total diet.

Mendall (1958) made a detailed study of food intake of ring-necked ducks in Maine and found that the tubers and seeds of two species of bulrush, seeds and vegetative parts of various pondweeds, and seeds of bur reeds (*Sparganium*) were major foods of adults, especially in spring and fall. Samples obtained during summer had a higher incidence of aquatic weeds and grasses, specifically wild rice (*Vallisneria*), followed by the seeds of spike rush (*Eleocharis*) and water lilies. Although nearly 90 percent of the adult food examined was of plant origin, samples from downy young contained about half animal matter, mostly aquatic insects. Plant materials included many of the same items taken by adults, even including the tubers of bulrushes. Mendall noted that ring-necked ducks generally feed in shallower waters than do other diving ducks in Maine, and they preferred sites less than five feet deep for foraging. They also tip-up at times and generally remain submerged for relatively short periods of about 8 to 25 seconds.

Sociality, densities, territoriality. Mendall (1958) stated that ring-necked ducks are not averse to nesting in close proximity to one another, sometimes nesting only 5 to 6 feet apart. One quarter-acre island was found to support six ring-necked duck nests and one black duck nest. This would suggest the possibility of fairly high nesting densities in favorable habitats. Jahn and Hunt (1964) indicated that a six-year average density of ring-neck pairs per 100 acres of wetlands in Wisconsin was nine pairs for the northern highland and six pairs for the central plain. Perhaps a more realistic measure of ring-neck densities is that provided by Lee et al. (1964b) for a 2.5-square-mile study area in Mahnomen County, Minnesota. In four years, the estimated population of ring-necked ducks ranged from 4.5 to 12 pairs per square mile and averaged 8.8, or almost twice as many as either the redhead or the canvasback populations. Mendall (1958) calculated maximum densities of various study areas as ranging from a pair per 6 to 23 acres, with the former density

Ring-necked duck, breeding pair swimming

apparently close to the maximum possible. He believed that the unusual small home range and low level of intraspecific aggressiveness accounted for this remarkably high potential breeding density.

Mendall (1958) has discussed the possible role of territoriality in ring-necked ducks and noted that defense of the female had often been seen, but defense of specific areas had been noted only a few times, and then only prior to or during nest site selection. He nevertheless accepted the concept of territoriality as applying to this species, assuming that a condition of mutual respect served to avoid friction between pairs. Yet, little or no evidence of territorial boundaries could be found, and Mendall was unable to explain how concepts of classic territoriality might be applied to this species.

Interspecific relationships. The rather specialized habitat preferences of the ring-necked duck largely place it out of direct contact with other pochards on the breeding grounds, and probably only the black duck regularly breeds in its preferred nesting habitats. On wintering grounds it most often associates with scaup but tends to occupy less brackish waters.

Predators of eggs include minks, crows, ravens, raccoons, foxes, skunks, and perhaps other species, but the first four probably account for the largest number of losses (Mendall, 1958). Ducklings have reportedly been taken by snapping turtles, minks, and foxes, and no doubt other predators also account for some losses.

General activity patterns and movements. Little specific information on daily activity rhythms and local movements are available. Mendall (1958) noted that ring-necks have a regular daily feeding pattern, except during courtship and the early stages of nesting. He mentioned that their morning foraging flights are seldom as early as those of black ducks and goldeneyes, but the evening feeding period is at about the same time.

Mendall also noted that on a 70-acre study area (Barn Meadow), the first pairs to arrive in spring initially had rather large home ranges ("territories"), which decreased in size as other pairs moved in. Up to seven pairs were found to occupy the marsh, and additional pairs might have had their nests within it but established waiting sites and/or "territories" elsewhere. Thus, it would seem that home ranges of this species might vary in size during the breeding season but in general are probably relatively small and localized.

Social and Sexual Behavior

Flocking behavior. Mendall (1958) stated that fall migrant flocks of ring-necks are generally larger than those in spring, but groups of 10 to 25 birds are frequent. During periods of mass migration large flocks might sometimes occur, but the usual flock size of groups arriving at the wintering grounds is 5 to 25 birds. Apparently there is a substantial segregation of the sexes during fall migration, although the details of this are still obscure.

Spring migrant flocks are usually rather small, with groups of about 6 to 30 being typical. The earliest migrants are usually pairs and courting groups, followed by many unpaired birds having a large excess of males (Mendall, 1958).

Pair-forming behavior. The pair-forming behavior of ring-necked ducks begins on the wintering grounds and probably reaches a peak during spring migration in March and April (Weller, 1965). By mid-May, when nesting is under way, it is seen very little, although scattered occurrences might take place until mid-June (Mendall, 1958).

The male pair-forming display of ring-necked ducks includes the usual pochard head-throw (Fig. 14G–H) and "kinked-neck" calls, both of which are associated with a soft whistling note, neck-stretching, a rudimentary head-forward or "sneak" posture, and a few other less conspicuous displays (Johnsgard, 1965; Mendall, 1958). The female's inciting movements and calls are much like those of other pochards and serve the same function. Marquardt (cited in Mendall, 1958) noted the importance of the female's inciting in stimulating and maintaining male display activity, and it is certainly true that inciting behavior seems to play a major role in pair formation. The response of the preferred male to such behavior is usually to swim

Ring-necked duck, breeding male swimming

beside or ahead of the female and turn-the-back-of-the-head toward her. Ripley (1963) described an unusual lateral threat display in males that has not been noted by other observers, but he evidently failed to observe some of the more typical ring-neck displays.

Copulatory behavior. Ring-necked ducks normally precede copulation with mutual bill-dipping and dorsal-preening behavior. The postcopulatory display is reportedly the usual male call and bill-down posture typical of all pochards so far observed (Johnsgard, 1965).

Nesting and brooding behavior. Mendall (1958) has provided a large amount of information on nesting behavior, part of which might be summarized here. Females apparently select the nest site but are accompanied by males. In early-nesting birds as much as a week or ten days might elapse between site selection and the laying of the first egg, whereas late-nesting birds might begin to lay almost immediately. Sometimes

Ring-necked duck, breeding males swimming

little or no actual nest is evident at the time the first egg or two are deposited, and until about the sixth egg there is still usually little nest shape evident. However, down is then usually added as the clutch is completed, and the vegetation overhead might be woven together to form an overhead arch. Ramps might be built to nests elevated above the ground surface, and runways to the nearest water are established.

Eggs are usually laid in the forenoon, during visits lasting 15 minutes to three hours. Incubation apparently begins on the day the last egg is laid. During early stages of incubation the female might spend considerable time away from the nest, especially on cool days, and the period of strongest incubation behavior is between 9:00 a.m. and 3:00 p.m. During the last two weeks of incubation, the females incubate more closely, and during this period the male usually abandons his mate.

Pipping of the egg occurs 24 to 48 hours prior to hatching, and most eggs hatch within a 6- to 8-hour period. The female then normally broods her young for at least 12 hours, and the family leaves the nest in late afternoon or, more frequently, shortly after sunrise on the day following hatching. In contrast to most other waterfowl, the female might bring her young back to the nest for brooding purposes for 2 to 4 days after hatching, or even longer. Further, few females abandon their broods prior to the time of their fledging, even when they themselves have become flightless. There are apparently few if any brood mergers in this species, and no apparent friction between the parents of broods feeding in close proximity.

Postbreeding behavior. Males begin their postnuptial molt even before they have abandoned their mates and soon begin to gather with other males that never attained mates or have abandoned theirs. There is probably a northward molt migration of birds that breed in Maine, but the distance and location of molting areas are still only poorly known. One such area, the Saint John River estuary of New Brunswick, regularly supports several hundred molting birds in August and early September. The duration of the flightless period is probably 3 to 4 weeks, with the females having their flightless stage about a month later than the males. As young birds attain the power of flight, they begin to wander about, forming loose flocks that seem to disperse in a haphazard fashion. Before long, however, cooling weather in fall brings on initial gatherings in preparation for the southward migration.

Tufted Duck
Aythya fuligula (Linnaeus) 1758

Other vernacular names. None in North America.

Range. Breeds in Iceland, the British Isles, and most of Europe and Asia north to 70° latitude and south to central Europe, the Balkan Peninsula, the Kirghizstan steppes, Lake Baikal, the Amur River, Sakhalin, and the Commander Islands. Winters from its breeding range south to northern Africa, the Nile River valley, the Persian Gulf, India, southern China, and the Philippines.

Subspecies. None recognized.

Measurements. *Folded wing:* Delacour (1959): Males 198–208 mm; females 189–202 mm. Owen (1977): Adult males, ave. 202.8 mm; adult females, ave. 195.9 mm.

Culmen: Delacour (1959): Males 38–42 mm; females 38–41 mm. Owen (1977): Adult males, ave. 39.5 mm; adult females, ave. 38.5 mm.

Weights (mass). Owen (1977): Adult males, ave. weight 711.4 g; adult females, ave. 673.2 g. Bauer and Glutz von Blotztheim (1969): 21 males (January), ave. 872 g, max. 1,020 g; 11 females, ave. 759 g, max. 955 g. Dementiev and Gladkov (1967): Males (February) 1,000–1,400 g (ave. 1,116 g); females, 1,000–1,150 g (ave. 1,050 g).

Identification

In the hand. Males in breeding plumage have a black head and neck with a long, narrow crest that nearly touches the back. The upperparts are blackish, the scapulars having a greenish cast and faint vermiculations. The breast, tail coverts, and tail are all black, but the abdomen and flanks are white. The upper wing-coverts are dark brown, and the secondaries are white with black tips. The primaries are dark brown, with the inner feathers having gray or white on the inner webs. The iris is yellow, the bill is pale blue with a black tip, and the legs and feet are lead blue with darker webs. The bill is slightly narrower and shorter than that of a scaup (maximum culmen length 42 mm; maximum width under 24 mm) and is only slightly wider toward the tip than at the base, but both the nail and adjacent tip are black in color. Whitish vermiculations are lacking on the back and upper wing-coverts of both sexes.

Adult males have a thin, drooping crest, which is rudimentary in females, but females lack a white cheek mark large enough to be continuous across the forehead, and some females have a whitish mark at the sides of the mandible. *Males in eclipse* are much like females but are more grayish throughout, with some

Fig. 17. Tufted duck, pair swimming

vermiculations showing on the flanks. *Females* are similar to scaup females but are darker dorsally, have a small occipital crest, and show little or no white at the base of the bill. Some females also exhibit a white area on the under tail-coverts, but most are dull brown in this region. The soft-part colors are similar to those of the male. Yearling females have a brownish iris color, and, if comparable to the lesser scaup, the iris color is likely to be transitional between brown and yellow by the second year. In adults the eye color of both sexes is yellow, but in females the iris color is less bright and the bill is more grayish. *Juveniles* are similar to adult females but have brown eyes; young males are somewhat vermiculated dorsally and have darker heads.

In the field. Tufted ducks are most likely to be confused with greater scaups, but the crest of the male will normally allow for separation, and even the female exhibits a slight crest. Females also have less white in front of the eyes than do scaup females, but this trait is not useful in summer, when female scaup acquire a brownish face pattern. In flight, tufted ducks are extremely difficult to separate from scaup, and the head characteristics just mentioned provide the best clues. The calls of the male tufted duck consist of a mellow *whee' oo* and a rather windy *wha'wa-whew*, which are very similar to the corresponding calls of the greater scaup. A low growling call and a *gack* or *quack* note are produced by females.

Tufted duck, breeding pair swimming

Age and Sex Criteria

Sex determination. The presence of a definite elongated crest or of definite vermiculations on the scapulars, sides, or flanks indicates a male. In eclipse plumage the sexes might be difficult to distinguish, but some vermiculations are present on the male's grayish sides and flanks, whereas in females the sides and flanks are more uniformly brownish. According to Veselovsky (1951), juvenile males can be distinguished from females by their darker brown head and neck color and bluish gray rather than dark brown bill. Kear (1970) found that by 35 days of age males have a brighter yellow eye color than females.

Age determination. Although the adult plumage is attained by the end of December, individuals carrying notched tail feathers have been taken as late as April (Kear, 1970). Bauer and Glutz von Blotztheim (1969) noted that the axillaries and greater and middle upper wing-coverts of immature birds are shorter and have more frayed edges than those of adults.

Tufted duck, breeding male swimming

Occurrence in North America

Either the tufted duck has become much more frequent in North America during recent years, or it was con-fused earlier with the somewhat similar ring-necked duck. The earliest known North American records were from Saint Paul and Attu Islands, Alaska. In recent decades there have been eBird records from at least 14 Attu locations and 7 Saint Paul locations. There are at least 12 eBird records from Adak Island, and one or more records from Kiska Island, Buldir Island, and Unalaska Island, as well as from several locations along Alaska's southern and western coasts and interior.

In Canada tufted ducks have been most frequently reported from British Columbia, the first sight record being obtained in 1961 (Godfrey, 1986). In western Canada eBird records also include the Yukon, North-west Territories, Alberta, and Saskatchewan. In eastern Canada, they include at least Ontario, Quebec, New-foundland, Nova Scotia, and Prince Edward Island (American Ornithologists' Union, 1998).

Pacific coast records from south of Canada are centered in Washington. In the Seattle area the species has been seen almost annually in recent years. There are at least 18 records from Oregon (Gilligan et al., 1994) and by 1993 there were 69 records from California, of which 22 were repeats (Patten, 1993).

Atlantic coast and eastern US records have become so numerous as to make a complete listing impossible. South of Canada the largest number of state wintering records are probably from Massachusetts, where the species was first reported in the 1950s. There are also eBird records for New England from Maine, New Hampshire, Vermont, Connecticut, and Rhode Island as well as New York and New Jersey.

There are also many documented state records from the continental interior, which include Arizona (American Ornithologists' Union, 1998), Wyoming (Faulkner, 2010), Montana (Marks, Hendricks, and Casey, 2016), Kansas (Thompson et al., 2011), and Nebraska (Johnsgard, 2012). There are also eBird records from Idaho, Colorado, Arkansas, Kentucky, Illinois, Ohio, Michigan, Pennsylvania, Maryland, Delaware, and North Carolina.

Greater Scaup
Aythya marila (Linnaeus) 1761

Other vernacular names. Big bluebill, bluebill, broadbill

Range. Breeds in Iceland, in northern Europe and Asia to northern Siberia, and in North America from arctic Alaska and arctic Canada east to the eastern shore of Hudson Bay, northern Labrador, Anticosti Island, and Newfoundland. In North America winters on the Pacific coast from the Aleutian Islands to California, on the Gulf coast almost to Mexico, on the Atlantic coast from Florida to southern Canada, and on the Great Lakes, especially on Lakes Erie and Ontario.

North American subspecies. *A. m. mariloides* (Vigors): North American Greater Scaup. Breeds in North America, as indicated above.

Measurements. (*Note:* Owen's data are for European *A. m. marila*; Delacour's data probably include *A. m. marila*.) *Folded wing:* Delacour (1959): Males 215–233 mm; females 210–220 mm. Owen (1977): Adult males, ave. 221.5 mm; adult females, ave. 212.0 mm.

 Culmen: Delacour (1959): Males 43–47 mm; females 41–46 mm. Owen (1977): Adult males, ave. 44 mm; adult females, ave. 42.8 mm.

Weights (mass). (*Note:* Owen's and Schiøler's data are for European *A. m. marila*.) Owen (1977): Adult males, ave. 711.4 g; adult females, ave. 673.2 g. Nelson and Martin (1953): 60 males, ave. 2.2 lb. (997 g); 43 females, ave. 2.0 lb. (907 g); max. (both sexes) 2.9 lb. (1,314 g). Schiøler (1926): 12 adult males, ave. 1,256 g; 8 immature males, ave. 1,131 g; 12 adult females, ave. 1,182 g; 7 immature females, ave. 1,024 g.

Identification

In the hand. As with the lesser scaup, the presence of a white speculum, a bluish bill that widens toward the tip, and vermiculated gray (breeding males) to brownish (females and juveniles) upperparts eliminate all other species of ducks. For distinction from lesser scaup, see the account of that species.

In the field. In good light, male greater scaup in breeding plumage exhibit a greenish rather than purplish gloss on the head and have a relatively low, uncrested head profile. Additionally, the back appears more grayish, since it has a more finely vermiculated pattern. In flight, the extension of the white speculum to several of the inner primary feathers might be apparent. Female greater scaup are difficult to distinguish from female lesser scaup unless they are seen together. Greater scaup are slightly larger and have more white on the face, especially on the forehead. The calls of the females of both species are similar, the most frequent one a low, growling *arrrr* that is somewhat weaker in the lesser scaup. The courtship calls of the male greater scaup are a very soft, cooing *wa'hoooo* and a weak and very fast whistle *week-week-week*, compared with the lesser's faint *whee-ooo* and a single-noted *whew* whistle (Johnsgard, 1963). In both species these calls might be heard only at fairly close range during courtship activity.

Age and Sex Criteria

Sex determination. Although both sexes might have vermiculated scapulars, those of males are predominantly white, whereas those of females are predominantly dark. Females always lack flecking on the tertials and usually also on the greater and middle coverts, whereas males usually exhibit flecking (Carney, 1964). Most males older than juveniles have black or blackish feathers on the head, breast, or rump, and might exhibit vermiculations on the flanks. Flank vermiculations are lacking in females.

Age determination. Juvenal tertials are usually frayed to a pointed tip, whereas those of adults have more rounded tips. Additionally, juveniles' tertials are rough, often narrower, and duller than those of adults. The tail should also be examined for squarish and notched-tipped feathers. Judging from lesser scaup information, females with yellow irises are at least two years old; some two-year-olds might have brownish yellow eyes, or brown eyes, as are typical of yearlings (Trauger, 1974).

Distribution and Habitat

Breeding distribution and habitat. In North America the greater scaup is primarily found considerably farther north than is true of the lesser scaup. In their choice of breeding habitats, the greater is more tundra-adapted, and the lesser is more closely associated with boreal forests.

In Alaska, the principal breeding range extends from the Alaska Peninsula northward along the coast of the Bering Sea to the valley of the Kobuk River (Gabrielson and Lincoln, 1959) and locally north to the Beaumont Sea coast. It is common on the Aleutian Islands during spring and summer months, and it has also been reported to nest on Amchitka Island (Kenyon, 1961). Breeding no doubt occurs over much of the interior of western and central Alaska, since scaup made up over a third of the ducks identified on 1960–69 aerial breeding-ground surveys made by the US Fish and Wildlife Service. However, at least in central and eastern Alaska, the lesser scaup also breeds, and the relative status of the two species across the state is still somewhat unsettled.

Greater scaups are known to nest in the Minto Lakes area of central Alaska, but the lesser scaup was judged more common than the greater scaup farther north around Anaktuvuk Pass (Irving, 1960), and at Old Crow (located near the Alaska-Yukon border) the same situation seems to apply (Kessel, Rocque, and Barclay, 2002). Likewise, King (1963) noted that in the upper Yukon River area, only 52 of more than 12,000 scaup banded while molting were greater scaup, and evidently only a few nest in that area. Kessel,

The breeding (hatched, with denser concentrations inked), wintering (shaded), and marginal (stippled) range of the greater scaup.

Rocque, and Barclay (2002) stated that, apart from the Minto Lakes area, most of the Alaskan greater scaup nest in the Seward Peninsula–Kotzebue Sound region and the Bristol Bay–Yukon-Kuskokwim Delta region.

In Canada the probable breeding range extends at least sporadically from the northern Yukon through the Northwest Territories (Great Slave Lake) and Alberta (Ministik Lake, Bowden Lake, and Red Deer River) eastward through the Mackenzie District of Northwest Territories and locally across northern Manitoba, the Hudson Bay coast of Ontario and Quebec, the Ungava Bay coast, central Labrador, and eastern Newfoundland (Kessel, Rocque, and Barclay, 2002).

In summary, the breeding habitat of the greater scaup is primarily that of tundra or low and sparse taiga scrub closely adjacent to tundra, rather than boreal forest. Hildén (1964) observed that this species requires relatively open landscape, cool temperatures, and shallow waters of high trophic quality with open, preferably grassy, shores. He noted a strong social attraction toward nesting gulls or terns, and he found the highest nesting abundance to occur on islets with grassy or herbaceous cover, lower use of islets dominated by boulders or rocks, and little or no use of gravel-covered or wooded islets.

Population. It is very difficult to judge the North American population of the greater scaup, since they are impossible to differentiate from lesser scaup during aerial inventories, and most of their remote breeding grounds are still uncertain or unknown. Kessel, Rocque, and Barclay stated (2002) that summer populations of greater scaup on their breeding grounds are best obtained by using coastal tundra surveys in Alaska. In surveys from 1978 to 2001 these routes averaged 430,000 birds and ranged from 340,000 to 642,000. Populations from Hudson Bay eastward were estimated at about 6,000 pairs.

Alaskan counts of tundra-breeding scaup since 1978 have trended slightly upward, but more inland scaup counts have been in a period of prolonged decline since the late 1970s. No doubt most of these birds would have been lesser scaup; the relatively small greater scaup population at Minto Lakes declined dramatically after the 1950s, judging from greater scaup to lesser scaup ratios.

Midwinter counts of scaup have some of the same accuracy problems as summer counts, but the average US estimate of greater scaup for the period 1961–2000 was 192,000, with declining estimates of from about 400,000 in the early 1960s to about 200,000 by 2000. Kessel, Rocque, and Barclay (2002) used greater scaup to lesser scaup harvest-ratio data from 1962 to 1999 to conclude that 66 percent of the North American greater scaup winter in the Atlantic Flyway, 14 percent in the Pacific Flyway, 19 percent in the Mississippi Flyway, and 1 percent in the Central Flyway. This result is rather surprising and might suggest that many more greater scaup breed in eastern Canada than is currently appreciated. My 1975 estimate of 750,000 greater scaup in North America might not have been far from the truth at that time; recently Baldassarre (2014) similarly estimated a stable North American population of 560,000–710,000. The combined-species national 2016 population was estimated at 5.0 million birds and considered stable (US Fish and Wildlife Service, 2016).

The size of the Eurasian population of greater scaup is also uncertain, but about 500,000 was suggested for the European *A. m. marila* population in the late 1990s, and 200,000 to 400,000 birds for the east Asian population, which is sometimes separated taxonomically as *A. m. mariloides* (Kear, 2005).

Fig. 18. Greater scaup, adult male swimming

Wintering distribution and habitat. In Alaska greater scaup winter commonly along the Aleutian Islands, Kodiak Island, and along the coastline of southeastern Alaska (Gabrielson and Lincoln, 1959). In Canada these birds regularly winter along the coast of British Columbia, on some of the Great Lakes, and along the Atlantic coast from southern Quebec eastward through the Maritime Provinces and Newfoundland (Godfrey, 1986).

South of Canada, western greater scaup might be found in winter along the coasts of Washington, Oregon, and California, southward to central California. There are occasional wintering birds farther south, but only rarely do they range as far as Mexico (Leopold, 1959).

On the Atlantic coast the greater scaup is most abundant along the coast of New England. Maximum numbers seen during the annual Audubon Christmas Bird Counts generally occur along coastal New York. To the south of New York, the relative abundance of greater scaup depends largely on the severity of the winter, with southern movements greatest in years of severest winters, so that in the Chesapeake Bay area either species might be more common during a particular year (Stewart, 1962). Stewart (1962) stated that in the Chesapeake Bay region the greater scaup is generally largely restricted to brackish and salt estuarine bays and coastal bays during winter, although some migrant birds use fresh and slightly brackish waters for brief periods.

Farther south in South Carolina and Georgia the greater scaup is quite rare (Sprunt and Chamberlain, 1949; Burleigh, 1958). Burleigh (1944) found only a single definite specimen from the Mississippi Gulf coast. Considering both the long migratory distance and the cold-weather tolerance of this species, it is apparent that the Gulf coast is not a part of its regular wintering range.

Note: In the half-century since the preceding description was written, there has been a gradual movement northward of nearly all North American winter bird populations, especially waterfowl (Johnsgard, 2015), often to the extent of several hundred miles.

General Biology

Age at maturity. Ferguson (1966) reported that 7 of 12 aviculturists found initial breeding of greater scaup in their second year of life, and only three reported first-year breeding. Comparable data on wild birds are limited, but all the females found breeding at Great Slave Lake were at least two years old (Weller et al., 1969), and only 1.4 percent of those found on Icelandic breeding grounds were yearlings (Bengtson, 1972b).

Pair-bond pattern. Greater scaup renew their pair-bonds on a yearly basis. In captivity, pair-forming behavior might be seen from late fall through winter and early spring, and probably the same applies to wild birds.

Nest location. Weller et al. (1969) noted that on the West Mirage Islands of Great Slave Lake, greater scaup typically place their nests in the grass of the previous year, often in rock cracks or near water. Of 29 nests, the average height above water level was 7 feet, whereas 28 nests averaged 19 feet away from the nearest water. Flint et al. (2006) stated that the 1,056 nests they found on the Yukon-Kuskokwim Delta were usually among low grasses and sedges along pond shorelines, and 81 percent of 30 greater scaup nests on the West Mirage Islands of Great Slave Lake were in grasses, sedges, or mixed grass–sedge vegetation (Weller, Trauger, and Krapu, 1969).

In a nesting study on Iceland, Bengtson (1970) analyzed the locations of 2,016 greater scaup nests. He found nearly twice as many nests per unit area on islands as on mainland habitats (331 vs. 180 nests per square kilometer). Favored nest sites were under the perennial herbaceous angelicas (*Angelica* and *Archangelica*) and shrubs, especially those under 0.5 meter high. Other herbaceous cover and sedges were used to a much lesser extent, and only one nest was found in a hole. Bengtson found that scaup exhibited a tendency for nesting in aggregated or clumped patterns and, in general, nested fairly close to water.

Clutch size. Weller et al. (1969) noted that 49 nests averaged 7.8 eggs but ranged up to 22 in number. Including only the 39 currently incubated clutches, and excluding those numbering in excess of 12 eggs as assumed multiple efforts, the average clutch was 8.5 eggs. Hildén (1964) reported an average of 9.68 eggs in 360 clutches, with a modal clutch size of 10 eggs and a maximum of 17. He also found brood parasitism to be prevalent, with both intraspecific and interspecific instances (in tufted duck and northern shoveler nests). Bengtson (1971) reported that 1,409 clutches of greater scaup in Iceland had an overall average of 9.73 eggs, although significant yearly differences in average clutch size (9.01–9.83 eggs) were present.

Greater scaup, breeding pair resting

Incubation period. The incubation period is generally estimated as 24 to 25 days (Hildén, 1964); Bengtson (1972) estimated a 26- to 28-day period.

Fledging period. The fledging period is reported to be 40 to 45 days (Hildén, 1964), which seems a remarkably short duration, considering the estimated 47- to 73-day fledging period of the lesser scaup, a 74-day period for the New Zealand scaup (Kear, 2005), and a 49- to 63-day period for the ring-necked duck. A 24-hour daylight period at the Arctic Circle would allow 24-hour foraging, but even the northern end of the Gulf lies south of the Arctic Circle, which would probably somewhat affect foraging opportunities.

Nest and egg losses. Hildén (1964) determined egg losses in 137 greater scaup nests in the Gulf of Bothnia, Finland. Of these 137 nests, 87 percent hatched, with crows and ravens accounting for most losses, and flooding causing a few. This relatively low loss to species such as crows might be the result of the high social attraction of greater scaup to nesting among gulls and terns, which tends to reduce crow depredations.

The effectiveness of gulls in reducing such predations is also greater late in their nesting season, when they are defending young, which might be advantageous to the late-nesting scaup. A more recent study by Bengtson (1972) confirmed the higher hatching success of scaup nests in gull or tern colonies than of nests not associated with them. The same was true for greater and lesser scaup nesting on islands among gulls and terns in Great Slave Lake (Fournier and Hines, 2001).

Brood parasitism in North America has been reported as being caused by lesser scaup, red-breasted merganser, and long-tailed duck, as well as cases of reciprocal intraspecific parasitism between greater and lesser scaup. Greater or lesser scaup eggs have also been found in the nests of gadwalls, northern pintails, and northern shovelers (Kessel, Rocque, and Barclay, 2002). In Iceland, Bengtson (1972) reported a 10.6 percent rate of intraspecific parasitism and an interspecific rate of 9.0 percent among 2,311 nests, with hatching success averaging lower among those presumably parasitized nests containing 12 to 14 eggs. Renesting occurred among about half of 41 females on the Yukon-Kuskokwim Delta when their eggs were taken early in the incubation period (Flint et al., 2006).

Juvenile mortality. Hildén (1964) found a tendency for scaup broods to intermix temporarily with those of tufted ducks, but he observed no indication of regular mergers of scaup broods. He did note one case of a female scaup with 15 young, which he thought might represent a merged brood. During three years of study he found that the rate of juvenile mortality ranged from 91 to 98 percent, and that much of the mortality was attributable to gull predation and bad weather. Since the young ducklings moved out into the open water of bays at an unusually early age, they were subjected to higher predation rates than were young tufted ducks and were also more likely to be caught in fishing nets.

Adult mortality. Boyd (1962) estimated an annual adult mortality rate of the Icelandic population of greater scaup as 48 percent. The annual survival of females from the Yukon-Kuskokwim Delta has been estimated at 79 percent, which is similar to Atlantic Flyway estimates of 75 percent for adult males and 70 percent for adult females (Kessel, Rocque, and Barclay, 2002).

General Ecology

Food and foraging. The summary of Martin et al. (1951) indicated that during winter and spring, a variety of animal materials (mollusks, insects, and crustaceans) seem to predominate in the diet, whereas 354 fall samples were predominantly composed of vegetable materials. The seeds and vegetative parts of pondweeds (*Potamogeton*), wild celery (*Vallisneria*), wigeon grass (*Ruppia*), and the vegetative parts of musk grass (*Chara*) were among the more prevalent plant materials found.

In a later study, Cronan (1957) analyzed the food contents of 119 greater scaup collected along the Connecticut coast of Long Island Sound between October and May. In this sample, animal materials constituted more than 90 percent of the total food volume, more than had been found in earlier studies. Cronan attributed this to the fact that all the birds were taken in coastal waters. He determined that the blue

Greater scaup, breeding male swimming

mussel (*Mytilus edulis*) was the most important single food by volume, although the dwarf surf clam (*Mulinia lateralis*) was of secondary importance both in volume and frequency of occurrence. Mollusks, most of which were bivalves, collectively made up nearly 90 percent of the total food contents. Cronan concluded that in different areas various mollusks serve as the primary foods, but the particular species utilized are evidently determined by their relative availability The English word "scaup" is a seventeenth-century variant of the Scottish *scalp*, a "mussel bed."

The only important plant food Cronan found was sea lettuce (*Ulva*), which is rapidly digested and probably was more important than the 3.6 percent of food volume that it constituted would indicate.

Cronan observed scaup of both species feeding during all daylight hours, with tidal stages being significant only where mollusk beds were exposed during low tide. Since the birds normally will not feed out of water, such low tides reduce foraging. Most foraging was in depths of less than 5 feet of water, but in one case diving in water 23 feet deep was seen. Temperature, water current, normal weather variations, wind, and cloud cover all had little or no effect on foraging, but human activities did strongly affect usage of local areas by scaup. Cottam (1939) noted that under conditions of human persecution, greater scaup often go to sea and return at night to the foraging areas, especially under moonlight.

Sociality, densities, territoriality. Bengtson (1970) noted that greater scaup exhibit a definite pattern of aggregation in their nesting distribution but did not know whether this was produced by social attraction or by some other environmental cause. On 13 areas he found an overall nesting density of 273 nests per square kilometer, or about one nest per acre. On islands the nesting density per square kilometer averaged 331 nests, and on the mainland 180 nests.

Interspecific relationships. In North America the lesser scaup is the nearest ecological counterpart of the greater scaup, and in Europe and Asia the tufted duck also occupies a similar ecological niche. Weller et al. (1969) found a considerable amount of brood parasitism between greater and lesser scaup, and Hildén (1964) likewise observed reciprocal brood parasitism between the greater scaup and the tufted duck in Finland. However, because of the rather generalized nest site requirements of these species, there appears to be little if any actual competition for nesting locations.

There is a good deal of similarity in the foods taken by lesser and greater scaup (Cronan, 1957; Stewart, 1962), at least when both are foraging in the same areas. Yet a sufficient degree of ecological segregation, apparently related to water salinity preferences and temperature tolerances, reduces such interactions to a fairly low level.

Although the presence of nesting terns or smaller gulls is highly attractive to scaup in providing nesting associates, at least larger species of gulls can be extremely destructive to ducklings during their first few weeks of life.

General activity patterns and movements. The observations of Cronan (1957) suggest that little obvious periodicity in foraging behavior can be detected in greater scaup, and since the birds are strictly open-water feeders, they do not undertake regular foraging flights to and from feeding grounds. Millais (1913) described foraging movements from the open sea to mussel beds at night, as well as at dawn and sunset. Dawson (1909) also noted there was a fall evening flight starting about half an hour before sunset from Drayton Harbor on the Washington coast, where the birds feed in shallow water and then move back out to sea.

Social and Sexual Behavior

Flocking behavior. The "rafting" behavior of migrant and wintering scaup flocks in large clustered groups is well known and is suggested by some of the species' colloquial names, such as "raft duck," "flock duck," and "troop duck." Scaup in rafts do not all forage at the same time; feeding and nonfeeding birds might be interspersed. When foraging in a water current, they often "drift feed," diving as they drift past a favored foraging area, and eventually flying back to the other end of the raft to begin drifting toward the foraging area

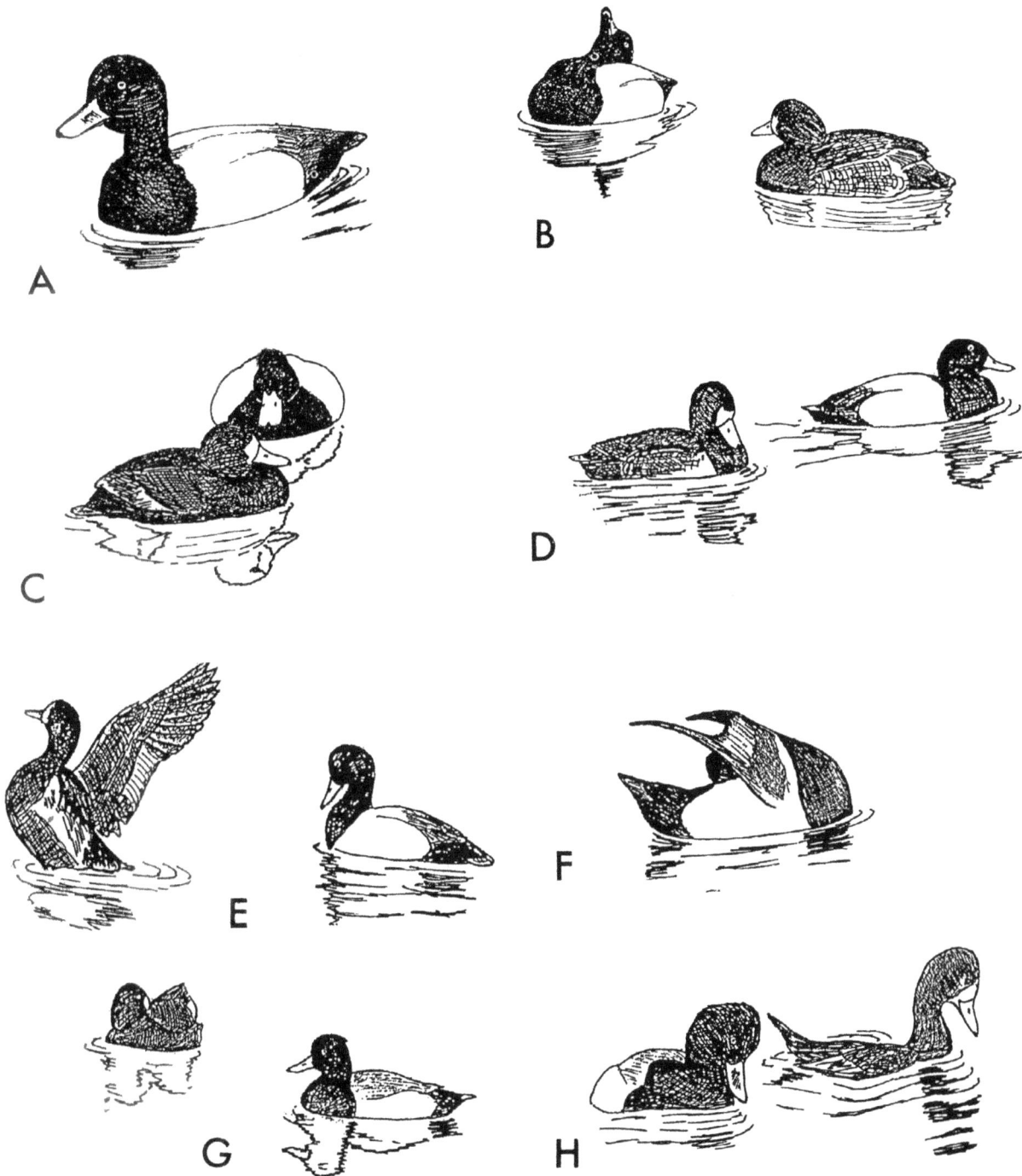

Fig. 19. Sexual behavior of greater scaup (A–E) and lesser scaup (F–H), including (A) kinked-neck call, (B) head-throw, (C) sneak posture, (D) male leading an inciting female, (E) female wing-flapping and male bill-down posture, (F) male mock-preening, (G) female preening-behind-the-wing, and (H) postcopulatory bill-down posture (after Johnsgard, 1965).

again (Cronan, 1957). Sizes of such rafts have not been extensively counted, but Audubon Christmas Bird Counts in the Long Island area have often shown in excess of 10,000 birds within a 15-mile-diameter circle.

Pair-forming behavior. The pair-forming behavior of the greater scaup is extremely similar to that of the lesser scaup, differing only in minor characteristics (Johnsgard, 1965). The inciting movements and calls of the females of these two species are virtually identical, and it is probable that some mixed courting groups might occur on common wintering grounds. However, wild hybrids between the two species are apparently unknown, although their recognition would prove to be extremely difficult.

Male pair-forming calls and postures include soft whistled "coughing" notes, uttered with an inconspicuous jerk of the wings and tail, and a very weak *wa'-hooo* note (kinked-neck call) that is produced during a head-throw display (Fig. 19B) or during slight neck-stretching. A sneak posture (Fig. 19C) and turning-the-back-of-the-head toward inciting females (Fig. 19D) are frequently performed, usually associated with a slight lowering of the crown feathers. Likewise, both sexes frequently perform a stereotyped preening-behind-the-wing toward the other (Fig. 19F), especially if the two are paired or in the process of forming pairs (Johnsgard, 1965).

Copulatory behavior. Copulation in greater scaup is usually preceded by the male bill-dipping, preening-dorsally, and preening-behind-the-wing. The female often responds with these same displays (which closely resemble normal comfort movements) and then assumes a prone posture. Following treading, the male typically releases the female's nape, utters a single call, and swims away from her in a rigid bill-down posture (Fig. 19E). The female might also assume this posture for a few seconds before she begins to bathe (Johnsgard, 1965).

Nesting and brooding behavior. Relatively little has been written on the nesting behavior of the greater scaup in North America. Hildén's (1964) study on the subarctic Gulf of Bothnia provides a good source of information. He found that the pair-bonds of this species last longer than in tufted ducks, on the average at least until the middle of the incubation period. The male remains near the nesting place and joins the female whenever she leaves the nest. In one case the male remained with his mate until hatching and was seen with the newly hatched brood.

Following hatching, young scaup ducklings feed mainly on the surface, catching floating insects or those flying just above the surface. Thus, the weather shortly after hatching, and its effects on insect abundance, as well as chilling effects on the ducklings, is critical to their survival. This is especially true of this species, which quickly leaves the shelter of the bulrushes and moves into the deeper water of the bays. There they are more directly exposed to the elements as well as to possible predation by gulls and perhaps also by predatory fish. Additionally, they must feed to a greater extent by diving because of the relative rarity of insect life. This demands more energy than does obtaining food from the surface or above it.

Postbreeding behavior. Hildén (1964) observed flocks of males as early as late June, about the time the first scaup broods were appearing. Most males were flocked by early July, when up to 50 birds were seen in a group. Except for a few that remained with apparently renesting females, the males then left the area and evidently molted elsewhere. Major molting areas in North America are still poorly known, but flocks have been seen on large tundra lakes of the Yukon-Kuskokwim Delta in mid-July (Kessel, Rocque, and Barclay, 2002), and Gabrielson and Lincoln (1959) mentioned that greater scaup are sometimes fairly numerous in southeastern Alaska during summer. Very possibly the coastal regions of the Northwest Territories or Hudson Bay also support molting scaup, although this is mere speculation.

Lesser Scaup
Aythya affinis (Eyton) 1838

Other vernacular names. Bluebill, broadbill, little bluebill

Range. Breeds from central Alaska east to Hudson Bay and Quebec to the St. Lawrence River valley, and southeast to Idaho, Colorado, Wyoming, and the Dakotas. Winters from British Columbia south along the Pacific coast to Mexico and Central America, rarely to Colombia, and along the Atlantic coast from the mid-Atlantic states south to Venezuela, as well as in Cuba, the Bahamas, and Lesser Antilles.

Subspecies. None recognized.

Measurements. *Folded wing:* (Delacour, 1959): Males 190–201 mm; females 185–198 mm. Kear (2005): 40 males, 193–226 mm, ave. 221.5 mm; 47 females, 184–205 mm, ave. 195.3 mm.
   Culmen: (Delacour, 1959): Males 38–42 mm; females 36–40 mm. Kear (2005): 40 males, 47–54 mm, ave. 50.0 mm; 47 females, 46–54 mm, ave. 49.7 mm.

Weights (mass). Nelson and Martin (1953): 130 males, ave. 1.9 lb. (861 g), max. 2.5 lb. (1,087 g); 144 females, ave. 1.7 lb. (770 g), max. 2.1 lb. (950 g). Combined data of Bellrose and Hawkins (1947) and of Jahn and Hunt (1964): 11 fall adult males, ave. 1.84 lb. (834 g); 36 immature males, ave. 1.74 lb. (789 g); 8 adult females, ave. 1.65 lb. (748 g); 36 immature females, ave. 1.76 lb. (798 g).

Identification

In the hand. Lesser scaup are best separated from greater scaup in the hand, and even then some specimens might remain doubtful. In *females*, the presence of a white facial mark and white on the outer webs of the secondaries exclude all species but the greater scaup. Female lesser scaup usually have no white on the inner webs of any primaries, although some might be quite pale. The length of the culmen in female lesser scaup is 36 to 40 mm, whereas female greater scaup have culmen lengths of 41 to 46 mm. Female lesser scaup rarely exceed 950 g, whereas female greater scaup average more than 950 g. *Males* can usually be distinguished from greater scaup by having (1) a purplish rather than greenish gloss on the head and a more extensive area of grayish vermiculations on the back, (2) no definite white on the vanes of the primaries (although the inner feathers might be quite pale), (3) a culmen length of 38 to 42 mm (vs. 43 to 47 mm for the greater scaup), and (4) a nail width of less than 7 mm (vs. at least 8 mm). The bill of the lesser scaup also tends to have a more concave culmen profile and to be relatively narrower at the base than that of the greater scaup.

The breeding (hatched, with denser concentrations inked), wintering (shaded), and marginal (stippled) range of the lesser scaup.

In the field. *Male* lesser scaup, when seen in sunlight, show a purplish gloss on the head and have a higher, more peaked crown profile, rather than a green-glossed head and a lower, more uniformly rounded crown profile. The mantle of the male lesser scaup also appears more speckled, since the vermiculations in these areas are coarser. In flight, the restricted amount of white on the secondaries might be evident, although the length of the white stripe on the secondary feathers is not definitive for species distinction. Lone *females* cannot be safely separated in the field, but lesser scaup females tend to show less white in front of the eyes than female greater scaup.

Age and Sex Criteria

Sex determination. Although the scapulars of both sexes might be vermiculated, those of females are predominantly dark, whereas those of males are predominantly white. Females have unmarked or only slightly flecked tertials, whereas males usually exhibit considerable flecking. The greater and middle coverts are usually unmarked in females and heavily flecked in males (Carney, 1964). The presence of blackish feathers on the breast or rump, vermiculations on the flanks, or head iridescence indicates a male.

Age determination. Juvenal tertials are usually frayed to a pointed tip, rather than being round-tipped, and the greater coverts tend to be narrower and duller than those of older birds. The tertials of immature birds usually lack flecking, but those of older males are flecked and of females are unmarked (Carney, 1964). Squarish tail feathers with notched tips indicate an immature bird. Trauger (1974) noted that the iris color of females gradually changes from olive brown in yearlings to yellow in three-year-old and older birds. Two-year-old birds cannot be distinguished from older birds in most cases, but none of 22 yearling females studied by Trauger had yellow eyes.

Distribution and Habitat

Breeding distribution and habitat. This strictly North American species of pochard has a fairly wide breeding range in both forest and grassland habitats. In Alaska it breeds commonly in the upper Yukon River valley (Yukon Flats) and the upper Tanana-Kuskokwim basin, including the Minto Flats (Gabrielson and Lincoln, 1959; Baldassarre, 2014).

In Canada lesser scaup breed southward from the treeline of the Yukon and Northwest Territories across the forested portions of British Columbia, Alberta, Saskatchewan, and Manitoba, east to Hudson Bay and Ontario. Farther east there are only spotty breeding records, such as in southeastern Ontario and western Quebec, with some breeding records east to the lower St. Lawrence valley (Godfrey, 1986; Austin, Custer, and Afton, 1996).

South of Canada, localized breeding occurs in eastern Washington (Yocom, 1951), eastern and southern Oregon (Gilligan et al., 1998), northern California (Reinecker and Anderson, 1960; Hunt and Anderson, 1966), central Arizona (Fleming, 1959), western Colorado (Kingery, 1998), and Wyoming (Oakleaf et al., 1962).

Fig. 20. Lesser scaup, adult male swimming

No documented breeding records exist for Nebraska; breeding is common in the northern Great Plains of northern and eastern Montana and the Dakotas, with the eastern limits of regular breeding occurring in northwestern Minnesota (Lee et al., 1964). There are also breeding records in Wisconsin (Jahn and Hunt, 1964), Ohio (Peterjohn and Rice, 1991), Indiana (Mumford, 1954), and Michigan, where the species has become much more common in recent decades (Brewer, McKeek, and Adams, 1991).

The preferred breeding habitat of lesser scaup consists of prairie marshes or potholes and lightly wooded parklands (Lee et al., 1964a). Godfrey (1985) characterized the Canadian breeding habitat as being the vicinity of interior lakes and ponds, with low islands and moist sedge meadows. Munro (1941) stated that nesting in British Columbia usually occurs around lakes of moderate depth, with bulrushes on shore and with brushy coves. Lakes with abundant amphipods and insect larvae support the best breeding populations.

Population. As noted in the greater scaup account, these two species are impossible to separate during aerial surveys, and their breeding ranges overlap somewhat in northwestern Canada. Bellrose (1980) suggested that the lesser scaup composes nearly 90 percent of North America's continental scaup population. Breeding waterfowl population surveys have indicated that more than two-thirds of all lesser scaup breed in the boreal forests of Canada and Alaska, a fourth breed in the prairie parklands of southern Canada and the northern United States, and less than 10 percent breed on the Alaskan tundra.

Between 1955 and 1995 these surveys indicated a breeding population of about 5.5 million, but with very large annual fluctuations. However, the long-term average (1975–95) has declined at a rate of about 125,000 birds per year, and there have been similar long-term (1955–95) declines evident in midwinter surveys (Austin, Custer, and Afton, 1996).

Wintering distribution and habitat. To a greater degree than any other pochard species in North America, the lesser scaup undertakes a notably long southward winter migration. A few lesser scaup winter in coastal British Columbia and on Lake Erie, but there is a general movement to saltwater regions of the southern United States and Mexico, and small numbers have migrated as far south as northern South America.

Midwinter inventories by the US Fish and Wildlife Service during the late 1960s indicated that nearly 90 percent of the scaup (both greater and lesser) wintered in the Mississippi and Atlantic Flyway states. From 2000 to 2010, the Midwinter Survey averaged 1.1 million scaup, with 36 percent in the Atlantic Flyway, 27 percent in the Mississippi Flyway, 21 percent in the Central Flyway, and 16 percent in the Pacific Flyway (Baldassarre, 2014).

In Mexico, lesser scaup have been second only to pintails in estimated numbers of wintering ducks, and are abundant along both coasts, the total averaging about 200,00 birds in 1961–2000 surveys. There have long been particularly large concentrations seen in Mexico on deep coastal lagoons of Nayarit, Chiapas, Veracruz, and Yucatan, although yearly variations in numbers and distribution were considerable (Leopold, 1959). From 1978 to 2006 the average number of lesser scaup inventoried on the east coast of Mexico was over 93,000, which were fairly equally distributed from the Laguna Madre Lagoon on the Texas-Tamaulipas border south to the Yucatan lagoons. Lesser scaup are also regular winter residents in Central America as far south as Panama (Wetmore, 1965). Some birds occasionally reach South America and, rarely, even Ecuador (Baldassarre, 2014).

Along the Atlantic coast, scaup winter from Newfoundland southward, but most of those occurring north of New Jersey consist of greater scaup. To the south the lesser scaup gradually increases proportionally so that in Florida it makes up nearly the entire wintering population (Addy, 1964). In that state lesser scaup winter mainly along the coast but also use some of the larger inland lakes (Chamberlain, 1960).

Stewart (1962) described lesser scaup habitat in Chesapeake Bay as consisting of fresh, slightly brackish, and brackish estuarine bays during migration, whereas brackish estuarine bays are the chief wintering habitat during most years. In severe weather scaup might move to salt estuarine bays as well. Unlike other ducks, their distribution was apparently not closely related to the distribution of aquatic food plants, a probable reflection of their greater dependence on animal foods.

General Biology

Age at maturity. In captivity, lesser scaup do not breed until they are two years old, according to 8 of 11 aviculturists responding to a survey by Ferguson (1966). There has been some speculation that a two-year period to maturity might also be typical of wild individuals as well (Munro, 1941), but evidence of breeding

Lesser scaup, breeding male swimming

by at least some female yearlings was found by McKnight and Buss (1962). It is probable, however, that only a small proportion of yearling birds successfully nest, as reported for both the lesser scaup (Trauger, 1971) and greater scaup (Weller et al., 1969).

Pair-bond pattern. Pairs are renewed each winter by lesser scaup, with pair-forming behavior beginning in January or February, which in general is more retarded than the pairings of redheads, canvasbacks, or ring-necked ducks (Weller, 1965). Pair-bonds are broken by the middle of the incubation period (Hochbaum, 1944).

Nest location. Munro (1941) noted that lesser scaup nests are built in dry situations under various kinds of cover, usually close to a lakeshore. Rienecker and Anderson (1960) stated they have a preference for nesting in dry uplands, with a slight tendency to choose islands with nettle (*Urtica*) cover. Vermeer (1968, 1970) noted a strong association in the nesting of lesser scaup with terns. Miller and Collins (1954) found that lesser scaup nest principally on islands, with grasses (50%), nettles (40%), and saltbush (10%) in descending order for cover preference. Miller and Collins found no nests over water, but all were 3 to 50 yards from water, usually in vegetation 13 to 24 inches high. Townsend (1966) noted that nearly 80 percent of the lesser scaup nests he found were in sedge cover, many of which were floating sedge mats. Keith (1961) reported that 198 lesser scaup nests in Alberta averaged closer to water (39 feet) than those of any of the surface-feeding ducks, and Townsend found that lesser scaup and ring-necked ducks were very similar in their placement of nests relative to water. Over half of the 40 nests that Gehrman (1951) found were within 15 feet of water.

Clutch size. Keith (1961) determined that the clutch sizes of lesser scaup decreased from 10.6 eggs early in the nesting season to 8.5 for late nests, with an overall average of 10.0 eggs in 131 nests. Likewise, Townsend (1966) found that 94 lesser scaup clutches averaged 9.0 eggs, with an average reduction of 1 egg per clutch for every 10.3 days delay in the seasonal onset of egg-laying.

Incubation period. Hochbaum (1944) determined a 22- to 23-day incubation period for lesser scaup eggs hatched in an artificial incubator, with a maximum of 26 days. Vermeer (1968) reported a 24.8-day average period for 18 clutches that were incubated by wild females.

Fledging period. Hochbaum (1944) indicated that captive-reared lesser scaup attained flight in 56 to 73 days. Lightbody and Ankney (1984) determined a very similar approximate 65-day fledging period for captive-raised birds, and range durations of 47 to 61 days have also been estimated (Schneider, 1965; Kear, 2005). This surprisingly broad span of estimated fledging periods (47 to 73 days) is notable.

Nest and egg losses. Nesting success rates seem to vary greatly by locality and year. High rates (60 percent or more) were reported by Miller and Collins (1954), Rienecker and Anderson (1960), and Townsend (1966). Much lower nesting success was estimated by Keith (1961), who found an overall 25 percent hatching success (higher on islands) in Alberta, and Rogers (1959), who noted a high incidence of nest losses to ground predators during a year of relative drought in Manitoba. Afton (1984) estimated the renesting incidence at 16.4 percent for 73 nests, but estimates tended to be lower for yearling birds and higher under improving water conditions.

Quite possibly the local availability of suitable nesting islands has a large effect on average hatching success of this species. Townsend found that lesser scaup nesting on islands had a higher hatching success than those nesting on the mainland. Vermeer (1968) noted that island-nesting lesser scaup had a high nesting success, whether or not nesting gulls were present.

Brood parasitism is common in both scaup species, both interspecifically and intraspecifically. In North Dakota, Lokemoen (1991) found that 48 percent of lesser scaup nests were parasitized on North Dakota island nests, mostly by redheads, but only 6 percent of the nests on peninsular sites were parasitized. Lesser scaup nests have also been parasitized by canvasbacks, white-winged scoters, red-breasted mergansers, and

Lesser scaup, breeding pair in flight

ruddy ducks (Austin, Custer, and Afton, 1996). Fournier and Hines (2001) found that of 65 lesser scaup nests 10 were parasitized by another lesser scaup, 17 by greater scaups, 5 by both lesser and greater scaups, and 1 by a northern pintail.

Juvenile mortality. Townsend (1966) calculated that the average hatch per nest of 55 lesser scaup clutches was 8.7 ducklings. Miller and Collins (1954) estimated a similar average hatch per clutch of 9.3 ducklings. Vermeer (1968) found a 100 percent duckling mortality in Alberta, mainly because of California gull predation. Afton (1984) estimated a 67.5 percent survival of broods during their first three weeks. Because of the prevalence of brood mergers in this species (Munro, 1941), counts of older-aged broods are not reliable measures of juvenile mortality.

Postfledging mortality rates for immature lesser scaup were calculated by Rotella, Clark, and Afton (2003), who estimated a survival rate of 55 percent for juveniles, as compared with 68 percent survival for adults, and 57 percent survival for females banded as adults. Smith (1964) estimated a 71 percent hunting-related mortality rate for immatures, versus 32 percent mortality for adult males.

Adult mortality. Longwell and Stotts (1959) calculated a 41.8 percent annual adult mortality rate for lesser scaups during the first six years following banding, or approximately the same rate as those they calculated for redheads and canvasbacks. More recently, Austin, Custer, and Afton (2000) stated that annual survival rate estimates have varied greatly, with most estimates ranging from 57 to 71 percent from year to year.

General Ecology

Food and foraging. In contrast to the three preceding pochard species, both species of scaup have higher rates of consumption of animal materials. Plant foods are similar to those of other pochards, including the seeds and vegetative parts of wild celery (*Vallisneria*), pondweeds (*Potamogeton*), wigeon grass (*Ruppia*), and various other submerged or emergent plants (Martin et al., 1951). Cottam (1939) found that animal foods, of which mollusks made up over half, constituted 40 percent by volume of samples from 1,051 lesser scaup sampled throughout the year. Insects were of secondary importance, and other animal foods constituted only about 3.5 percent.

More recently, Rogers and Korschgen (1966) analyzed 164 samples from adults on the breeding grounds, migration routes, and wintering grounds. Animal foods totaled 91.1 percent of breeding-ground food samples, 93.5 percent of fall samples, and 63.7 percent of winter samples. The most important foods were amphipod crustaceans from breeding area samples, mollusks from fall concentration areas, and fishes from wintering grounds. Harmon (1962) also noted the importance of animal foods, especially mollusks, from lesser scaup wintering areas in Louisiana.

Scaup spring and summer foods have been studied by several researchers, and most have commented on the importance of amphipods ("scuds") at this time of year. Munro (1941) noted this in British Columbia, as did Dirschl (1969) in Saskatchewan, Bartonek and Hickey (1969) in Manitoba, and Bartonek and Murdy (1970) in the Northwest Territories. In the last-named study, amphipods averaged more than half the total food volume among 35 scaup that had eaten them. Juveniles had consumed almost no plant materials but utilized free-swimming and bottom-dwelling organisms in wetlands averaging 3.5 to 4.0 feet in depth. Adults also usually foraged in water that was fairly shallow, but at times they fed at locations 15 to 20 feet deep. At such depths they would remain under water for about one minute, but in waters 8 to 10 feet deep they usually were submerged for less than half that duration (Cottam, 1939).

Sociality, densities, territoriality. Perhaps because of their land-nesting preferences and tendencies to nest on islands wherever these are available, lesser scaup often exhibit fairly high breeding densities and develop considerable sociality in nesting. Rogers (1959) reported that, on a square-mile study area in Manitoba, the breeding populations of lesser scaup totaled 51 and 65 pairs during two consecutive years. Vermeer (1970) also estimated a similar high nest density of 8 nests per 100 acres (51 nests per square mile) on islands in Lake Newell, Alberta. He later (1968) found that during two consecutive years there were 67 and 66 lesser scaup nests on two islands totaling 11 acres on Lake Miquelon, Alberta, representing an extremely high density of about 6 nests per acre. It is thus clear that territoriality must not play any significant role in determining nest dispersion in this species.

Interspecific relationships. The close relationship of lesser and greater scaup opens the possibility of interspecific competition and hybridization between these species. Their nesting areas overlap widely, and there is some evidence of interspecific conflict over nesting sites and parasitic egg-laying between the two species (Weller et al., 1969). Some reputed greater × lesser scaup hybrids have been described on Internet postings, but hard evidence for this seemingly likely cross is apparently still lacking.

The extent to which there might be food competition among the two scaup species is uncertain, but differences in migration routes and major wintering grounds tend to reduce contact between them. Stewart (1962) noted that in the Chesapeake Bay region, where both species winter, greater scaup were mostly restricted to brackish and salt estuarine bays and coastal bays, whereas lesser scaup ranged farther toward the upper limits of the adjoining estuaries, and moved out into salt estuarine bays only during unusually severe weather. Among birds collected in brackish and salt estuarine bays, both species had consumed the same gastropod mollusks (*Mulina lateralis, Brachiodontes recurvus*) in quantity, whereas samples of both species from salt estuaries included quantities of eelgrass (*Zostera*) and other gastropods (*Bittium, Mitrella lunata*). Thus, it appears that potential food competition between the two species is present, and probably the stronger tendencies of lesser scaup to use less salty or brackish waters, interior lakes, and more southerly wintering areas are the likely bases for reducing actual interspecific food competition.

General activity patterns and movements. Little specific information on daily activity rhythms and local movements is available. Phillips (1925) mentioned that the lesser scaup is primarily a daytime feeder but also forages to some extent at night. This species prefers to forage in shallow waters 3 to 8 feet deep and probably is less affected in its feeding rhythms by tidal fluctuations than are more marine species, such as the greater scaup.

Studies of local movements and home ranges on the breeding grounds are still inadequate. Hochbaum (1944) commented that the size of a lesser scaup's "territory" could be as large as an acre of open bay or as small as 40 yards of a narrow roadside ditch. He noted that paired birds sometimes make leisurely flights beyond the limits of their territory during evening hours, apparently for exercise.

Social and Sexual Behavior

Flocking behavior. The "rafting" behavior of scaup on their foraging areas is well known and often results in the concentration of large numbers of birds in localized areas. In Florida, where the lesser scaup is perhaps the most abundant wintering duck, large flocks of up to at least 200,000 birds have been seen around such areas as St. Petersburg, Fort Meyers, and Cocoa. If these birds were limited to the tidal estuaries that made up 15 percent of the survey area, they were spread out over no more than 25 square miles, averaging about 8,000 birds per square mile. In Louisiana the wintering scaup population is even higher; sometimes winter flocks totaling more than a million birds can be found on Lake Borgne, Lake Pontchartrain, and other lakes near New Orleans (Hawkins, 1964; Kinney, 2004).

Pair-forming behavior. Pair formation evidently begins fairly late on the wintering grounds (Weller, 1965) but during spring migration becomes relatively prevalent. In eastern and central Washington it might be seen from the time the birds first arrive in March to late April, by which time most females are paired (Johnsgard, 1955; Gehrman, 1951). Pair-forming displays of lesser scaup are very much like those of greater scaup and the other Aythyini species. The most elaborate posture is an extremely rapid head-throw, associated with a soft *whee-ooo* call. A sharper whistled note is also uttered during a convulsive cough-like movement, and a rudimentary form of the canvasback's crouched "sneak" display is sometimes performed.

During female inciting, which serves as the focal point of male courtship activity, the male often swims rapidly ahead and directs a turning-of-the-back-of-the-head display toward the female, simultaneously lowering the crown feathers to produce a distinctive low-headed appearance. To a larger extent than the other North American pochards (except perhaps the greater scaup), a ritualized preening of the secondary wing feathers that exposes the white speculum is prevalent during pair-forming display (Fig. 19G). Chases of the female, both in the air and under water, are fairly frequent (Gehrman, 1951), but it is uncertain whether these are courting chases or attempted rape chases.

Copulatory behavior. The precopulatory displays of the male consist of bill-dipping, dorsal-preening, and preening-behind-the-wing, which are sometimes reciprocated by the female. After treading is completed the male releases the female from his grasp, probably calls, and then swims away from his mate in a rigid posture with the bill pointed sharply downward (Fig. 19H) (Johnsgard, 1965).

Nesting and brooding behavior. The lesser scaup is one of the latest of the prairie-nesting ducks to begin nest-building and egg-laying, although any possible advantage of such late nesting remains obscure. Not only does predation intensity tend to increase late in the season but also renesting opportunities are reduced (Rogers, 1964).

The length of time required for nest building apparently has not been reported, but it is probable that eggs are laid on a daily basis. The male normally deserts the female when incubation begins (Oring, 1964), although he sometimes remains nearby as late as the middle of incubation (Hochbaum, 1944).

Following hatching, the brood is led to water, and brood rearing occurs in the relatively open water of large marshes. Females normally take good care of their young and usually feign injury when their broods are endangered. Frequently two females will jointly care for their merged broods and, when threatened, one will remain behind to threaten or feign injury while the other leads the combined brood to safety (Munro, 1941). Feigning injury and other defensive behavior decreases as the season progresses, and late in the season the females might simply attempt avoidance rather than defend their young (Hochbaum, 1944).

Postbreeding behavior. In the lesser scaup and related pochard species a relatively long period might elapse between the time the male abandons his mate and when he finally becomes flightless. Oring (1964) stated that this period might be as long as six weeks in the lesser scaup and the redhead. During this time the males gather in groups in favored areas. Hochbaum (1944) noted that male lesser scaup and redheads gathered in bands, moving from the Delta, Manitoba, marsh to the adjacent lake every morning and evening from mid-June through July. Later in July they spent more of the daylight hours on the lake, and finally remained there to undergo their wing molt in late July and August.

Farther north, the Great Slave Lake region is a favored molting area for thousands of lesser scaup, as is the Yellowknife region. Bailey (1983) stated that at least 95 percent of the lesser scaup nesting in the southern boreal forest staged or molted in the Saskatchewan Plain, on sparsely vegetated boreal lakes. He reported that as many as about 300,000 birds were present for a time at Macallum Lake, with smaller numbers at Garson, Gordon Beupré, and Kazan Lakes.

Tribe Oxyurini (Stiff-tailed Ducks)

This unique group of diving ducks differs from the rest of the Anatidae in so many respects that by any standard it deserves special attention. Of the eight species that are presently recognized, most are placed in the genus *Oxyura*, which means "sharp-tailed" and refers to the stiffened, elongated tail feathers typical of the group. In these species the tail feathers extend well beyond the unusually short tail-coverts and are numerous (18 or more) but relatively narrow, so that the individual rectrices tend to separate when spread. A double-annual molt of the rectrices is typical of stifftails, probably because of their importance in underwater maneuvering. There are also records of double annual wing molts in *Oxyura* and *Biziura*.

Likewise, their toes are relatively long. The long middle and outer toes result in associated web areas that are unusually large, resulting in strong propulsive abilities. Their legs are placed farther to the rear of the body than in any other waterfowl tribe, increasing propulsive efficiency during underwater swimming and diving. However, the legs' rearward position renders the birds unable to walk easily on land. Only two species of stifftails have been documented from North America, but South America claims four species (five if *ferruginea* is specifically recognized).

In the typical stifftails the bill is rather short, broad, and distinctly flattened toward the tip, and virtually all the foraging is done under water. At least in the North American ruddy duck, most of the foods taken are of animal origin, but the masked duck might have a more catholic diet. Nests of typical stifftails are built above water, of reed mats or similar vegetation, and often a ramp leads from the nest cup to the water, providing easy access. The birds are fairly heavy-bodied and have relatively short wings, so that flight is attained with difficulty in most species.

The masked duck is something of an exception to typical stifftail traits; its combination of small body size and a unique takeoff method allows it to land and take off with surprising agility from water of moderate depth. Although the species has most often been placed in a genus (*Nomonyx*) adjacent to *Oxyura*. I have previously concluded that its relationship, although clearly strained, is tentatively close enough to the ruddy duck and other *Oxyura* species to be retained within that genus. The ruddy duck is much more widespread and abundant in northern latitudes (breeding locally to central Alaska) than is the masked duck, which only occasionally crosses from the Mexican border into southern Texas as a breeding species.

The relationships of the other stiff-tailed ducks are still controversial. Until I deduced that the white-backed duck (*Thalassornis leuconotus*) is most closely related to the whistling-ducks rather than being an aberrant stifftail (Johnsgard, 1967), it was included as a member of the stiff-tailed group. I also concluded (1965) that the South American black-headed duck (*Heteronetta atricapilla*) is a transitional form between stiff-tailed and dabbling ducks, and is of special interest because it is unique among the Anatidae in being an obligate brood parasite. Lastly, the Australian musk duck (*Biziura lobata*) is unique in many behavioral and morphological ways, and it has some unique genetic features (Callaghan, Kear, and McCracken, 2005), but most of the male musk duck's strange traits can be explained as an example of sexual selection favoring

128

intense male-to-male competition for mates, in addition to strong selection for intersexual attraction signals having powerful visual, acoustic, and probably also olfactory stimuli for attracting females (Johnsgard, 1965, 1966).

Among the typical stifftails (*Oxyura*), the traditional taxonomic treatment has consisted of grouping the southern hemisphere species (maccoa duck, Australian blue-billed duck, Andean ruddy duck, and Argentine ruddy duck) distinct from the two northern species (North American ruddy duck and white-headed duck). On the basis of the many sexual behavioral differences distinguishing these two groups, I judged this grouping to be an incorrect association (Johnsgard 1965a, 1966, 1968, 1979; Johnsgard and Carbonell, 1996; etc.) and also concluded that the North American ruddy duck plus the two Andean ruddy duck populations (*andina* and *ferruginea*) are best considered conspecific with *jamaicensis*.

This general evolutionary/taxonomic scenario has recently been largely supported by the molecular studies of McCracken and Sorenson (2004) and the synopsis by Hughes (2005). Furthermore, based on their sexual behaviors, the Australian blue-billed duck and maccoa duck are apparently close relatives, whereas the Eurasian white-headed duck is seemingly only distantly associated with the other *Oxyura*, although Livezey (1995) placed it closest to *maccoa* within that genus.

Masked Duck
Oxyura dominica (Linnaeus) 1766

Other vernacular names. St. Domingo duck, squat duck, white-winged lake duck

Range. Breeds from coastal Texas (rarely) southward through Mexico, probably breeding mainly along the Gulf coast from Tamaulipas to Yucatan, and possibly also in western Mexico (reported from Sinaloa to Colima and Jalisco). Also resident in the Greater Antilles (uncommon in Cuba and Puerto Rico, apparently rare in Hispaniola) and the Lesser Antilles (breeding at least in Martinique and St. Lucia). Resident in Central America from Costa Rica to Panama but evidently only local in Guatemala, El Salvador, and Honduras. Also occurs throughout the lowlands of South America east of the Andes, from Colombia to northern Argentina. Probably nonmigratory in most tropical areas; seasonal movements are unstudied, but irregular appearances in the southern United States might be the result of periodic tropical storms.

Subspecies. None recognized.

Measurements. *Folded wing:* Johnsgard and Carbonell (1996): Males 136.2–138 mm (ave. of 5, 137.2 mm); females 132–139 mm (ave. of 5, 136.4 mm). Gomez-Dallmeier and Cringan (1989): Ave. of 39 adult males, 141 mm; ave. of 46 adult females, 139 mm.

Culmen: Johnsgard and Carbonell (1996): Males 31.1–33.9 mm (ave. of 5, 32.0 mm); females 30.8–34.2 mm (ave. of 5, 32.0 mm).

Weights (mass). Johnsgard and Carbonell (1996): Males 359–449 g (ave. of 19, 385 g); females 275–445 g, (ave. of 17, 346 g). Haverschmidt (1968): 9 males, 369–449 g (ave. 406 g); 6 females, 298–335 g (ave. 339 g). Gomez-Dallmeier and Cringan (1989): Ave. of 29 adult males, 372 g; ave. of 34 adult females, 358 g.

Identification

In the hand. This smallest of stiff-tailed ducks might be confused only with the ruddy duck, from which it can be easily distinguished by its white wing speculum; shorter bill (culmen under 35 mm), which does not widen appreciably toward the tip; and longer tail (at least 80 mm). Unlike the ruddy duck, the bill nail is large and not recurved below, the outer toes are shorter rather than longer than the middle toes, and the secondary feathers are long enough to hide the relatively short primaries when the wing is folded.

In the field. Although notably unwary, masked ducks are usually extremely difficult to find in the wild because they usually inhabit marshes that are extensively overgrown with floating and emergent vegetation, among which the birds mostly remain and under which they sometimes hide. The male in nuptial plumage is unmistakable with its black "mask," a long and sometimes slightly cocked tail, and a spotted cinnamon and brown plumage. However, most observations in the United States have been of females or female-like males; the male's breeding plumage apparently is held only from midsummer until early fall, the documented US breeding season.

The best field mark of females, immatures, and nonbreeding males is the strongly striped facial pattern, which consists of three instead of two buffy areas, including a superciliary stripe, an upper cheek stripe, and a buffy cheek and throat area. They are remarkably similar to female ruddy ducks and require considerable care in identification. Female ruddy ducks have only two buffy areas and completely lack a pale stripe above the eye. The species' diagnostic white wing markings are not visible unless the bird flies or flaps its wings, both of which are infrequent. When the birds take flight they rarely fly high but instead almost skim the marsh vegetation, suddenly slowing and dropping vertically downward out of sight into the marsh.

Age and Sex Criteria

Sex determination. Males in breeding (alternate) plumage might be readily recognized by their black "mask" and spotted rusty-cinnamon body color. Like the North American ruddy duck, this plumage alternates seasonally with a long-lasting basic, or "winter," plumage. Immatures and nonbreeding males have an overall plumage pattern extremely similar to that of females, and internal examination might be needed to determine sex.

Age determination. No information is available on the rate at which the adult plumage is attained. The juvenal plumage so closely resembles that of the adult female that probably the only certain plumage criterion of immaturity is the presence of juvenal tail feathers. Like the ruddy duck, these rectrices usually have conspicuous bare shaft tips with terminal enlargements marking the point where the downy tail feathers have broken off. Ripley and Watson (1956) stated that as compared with adults, subadults have noticeably wider and paler margins on the back and wing coverts, and almost downlike feathers on the underparts, which produce a rather mottled effect.

Distribution and Habitat

Breeding distribution and habitat. The North American distribution of this little-studied species is limited to Mexico and the coastal plain of southern Texas (with 94 documented Texas records as of 2014), where it is rare and irregular in occurrence (Lockwood and Freeman, 2014). Nonbreeding records also include many sightings in Florida (Robertson and Woolfenden, 1992), with at least 30 records as of 1994 (Bowman, 1995), and Louisiana, with at least 9 reports as of 1974 (Eitniear, 1999). There are also scattered records from at least eight other mostly eastern states, extending north as far as Massachusetts (American Ornithologists' Union, 1998). The Florida reports include one undocumented sighting of a brood at Loxahatchee National Wildlife Refuge in February 1977 (Bowman, 1995).

In the United States, the masked duck has long been known to have bred sporadically in the coastal plain of Texas, with records from Chambers County in 1930 (adult and five young), 1934 (female and one young), 1937 (three adults and five young), and 1967 (see below); Cameron County in 1937 (adult and five young); Brooks County, 1968 (female and six eggs); and San Patricio County (1968, pair with fledged young) (Oberholser, 1974).

In 1967 a well-documented US breeding record was obtained at the Anahuac National Wildlife Refuge, Chambers County, when a female and four young were found on October 2. Three adults were seen in the vicinity. In the autumn of 1968 a second nesting record was established when a nest with six eggs was found near Falfurrias, Brooks County, Texas (Johnsgard and Hagemeyer, 1969). A male and three females or immatures were also seen in November 1968 at Flour Bluff, Texas, and a pair was believed to have nested that fall in the vicinity of the Welder Wildlife Refuge, Sinton.

By 1969, the masked duck was seemingly established in southern Texas. Five or six pairs were at the Welder Wildlife Refuge that summer, the species was seen at Rockport until mid-July, and it again appeared at Anahuac National Wildlife Refuge in midsummer. No nests were found at Anahuac, but a female and eight young were observed there in late October 1969 (Johnsgard and Hagemeyer, 1969). No definite records of breeding in Texas were obtained in 1970 or 1971, but the occurrence of a hurricane there in early August 1970 caused extensive damage to coastal habitats of southern Texas and might well have affected nesting success. Later studies indicated Anahuac had additional nestings in 1992, 1993, and 1994 (Eitniear, 1999). In 1994 a group of 14 adults with ten young was seen at Attwater National Wildlife Refuge, Colorado County, between July and December 1994. Another nesting in Live Oak County was seen in 2007, when a pair and eight ducklings were found on a pond in late October. They were observed and photographed until at least mid-November (Eitniear, 2010).

As of 2016, Texas breeding efforts have come from at least Brazoria, Brooks, Cameron, Colorado, Hidalgo, Live Oak, Jefferson, and San Patricio Counties. The species has been reported every month in Texas but observed most frequently in April and December. Of the 94 documented occurrences in Texas, only 5 have been away from the coastal prairies. The last major incursion occurred in the early 1990s, with as many as 37 individuals seen on a lake in San Patricio County during February 1993 (Lockwood and Freeman, 2014).

Eitniear and Colón (2005) reported on a survey of masked ducks in the Caribbean region, with the largest number reported from Puerto Rico. There the birds are present throughout the year, with the largest number of sightings from February to June, after the rainy season. Second to Puerto Rico in number of sightings was Guadeloupe; birds were also seen in Antigua, Barbados, Grenada, Jamaica, and Trinidad. Most of the wetlands where masked ducks were seen were relatively small (Eitniear, 2012). The largest known flock reported from the Caribbean was a group of about 140 birds at Laguna Cartagena National Wildlife Refuge near Lajas, Puerto Rico, in February 2012. This refuge is a shallow precipitation-dependent wetland that is mostly covered by cattails and has a mat of floating vegetation. It is considered a critical habitat area for the endemic and endangered yellow-shouldered blackbird (*Agelaius xanthomus*), and the generally rare West Indian whistling-duck is a common resident.

The residential range of the masked duck, with selected US sight records (open circles), specimen records (inked circles), and breeding records (arrowheads).

In Mexico, the masked duck has been seen on freshwater marshes of Nayarit, Jalisco, Colima, Tamaulipas, and Veracruz, but there are very few definite Mexican breeding records (Howell and Webb, 1995). However, Berrett (1962) saw several adults and a single young bird on a roadside pond near Villahermosa, Tabasco, and no doubt breeding also occurs elsewhere along the Caribbean coast, and probably elsewhere in Mexico.

The breeding habitat of this species consists of tropical-like marshes or swamps densely vegetated with emergent vegetation and usually having lily pads, water hyacinths, or other floating-leaf aquatic plants extensively covering the water surface. The birds are sometimes seen in mangrove swamps, but most of the few breeding records are from freshwater habitats. In Texas the birds have often been found in marshy locations where yellow lotus (*Nelumbo lutea*), yellow water lily (*Nuphar mexicana*), giant bulrush (*Scirpus californicus*), and common cattail (*Typha latifolia*) occur (Blankenship and Anderson, 1993).

Population. Anderson (1999) estimated population densities on about 7,000 hectares (17,224 acres) of coastal Texas wetlands, finding that "lacustrine littoral aquatic-bed vascular wetlands" had the highest masked duck densities of 0.93 bird per hectare (2.1 birds per acre). This compares with his estimates of 0.4 bird per hectare on "lacustrine littoral aquatic-bed floating vascular wetlands" dominated by water hyacinth, and 0.15 bird per hectare on "palustrine scrub-shrub broadleaf deciduous wetlands." The ducks were seen on wetlands averaging 8.25 hectares (20.3 acres) in area, although most accounts suggest that small, shallow, and seasonal wetlands are more generally favored habitats.

During surveys of 652 160-acre wetland sites that Anderson, Muehl, and Tacha (1998) sampled from fall to spring during 1992 and 1993, a total of 47 masked ducks were counted. Extrapolating these data to the entire Texas coastal habitat, they estimated a total Texas population of about 3,800 birds, a figure that seems questionably high, given the rarity of sightings. Masked ducks are apparently much more common in rice-growing areas of Cuba, Venezuela, and Argentina, where up to 80 to 100 birds have been seen on a single marsh (Todd, 1996).

Wintering distribution and habitat. Seasonal movements of this species have not been studied but are probably not great. The birds have been seen during Audubon Christmas Bird Counts at the Welder Wildlife Refuge, where they also regularly have occurred during the breeding season. Habitats used in winter do not seem to differ from those used for breeding. Local or more extensive movements might be the result of regional changes in water levels.

General Biology

Age at maturity. Age at maturity is not yet documented, but maturity presumably is attained within one year, as is thought to be true of most other *Oxyura*.

Pair-bond pattern. Apparent pair-forming behavior has been seen during spring, but the overall pattern is uncertain. If pair-bonding occurs, it is probably of short duration (Johnsgard and Carbonell, 1996; Hughes,

Fig. 21. Masked duck, adult male standing

2005). Broods in Texas have been seen from midsummer to late fall being led by one or more females or female-like birds without any obvious males nearby.

Nest location. A number of clutches have been found in the rice plantations of Cuba; Bond (1961) noted that one of these that he observed was in a deep cup of rice stems and was placed just above the water level. Nests have been found in Panama among rushes and are said to lack down (Phillips, 1926). However, Dale Crider (pers. comm.) found that "snowflake" down was typical of the nests he found in Argentina.

Dale Crider (pers. comm.) reported that nests in northern Argentina were located in flooded rice fields among rice clumps beside deep water, into which the female could readily escape. The nests were roofed over and basketball-like in shape with lateral entries.

Clutch size. The usual clutch size is probably 4 to 6 eggs (Phillips, 1926; Wetmore, 1965). However, Bond (1961) listed eight Cuban clutches containing 8 to 18 eggs, which strongly suggests that dump-nesting (or brood parasitism) might occur in areas of nesting concentrations. Dale Crider (pers. comm.) also found a high average clutch size of about 10 eggs in Argentina, but obvious dump-nesting often made the clutches larger, with one nest of 27 eggs observed. According to Crider, the normal egg-laying rate is 1 egg per day.

Incubation and fledging periods. Dale Crider (pers. comm.) judged the incubation period of wild birds to be about 28 days, which would be an unusually long incubation period, considering the usual 22- to 26-day periods typical of the other similar-sized *Oxyura* species. Eitniear (1999) suggested that a 23- to 24-day period is more likely.

The fledging period is uncertain but probably at least 45 days (Eitniear, 1999) and might be as long as seven weeks, since Siegfried (1973b) reported that captive ruddy ducks fledged at seven to eight weeks.

Nest and egg losses. No specific information. Dale Crider (pers. comm.) found that the common caracara (*Caracara plancus*) was a major egg predator of most ducks in northern Argentina, and introduced Indian mongooses (*Herpactes javanicus*) have been implicated as predators in Cuba (Barbour, 1923).

Juvenile and adult mortality. No specific information regarding mortality is available. Dale Crider (pers. comm.) saw only rather small brood sizes and judged that the piranha was probably the most serious enemy of ducklings of this and other waterfowl species in northern Argentina.

General Ecology

Food and foraging. Phillips (1926) described foods found in Cuban specimens, which largely consisted of the seeds of smartweed (*Polygonum*), as well as small amounts of water lily (*Castalia*), rush (*Fimbristylis*), dodder (*Cuscuta*), and saw grass (*Cladium*). Three Cuban specimens mentioned by Cottam (1939) had virtually the same contents and no doubt represented the same specimens. Weller (1968) noted several types of seeds in the gizzard of a male. Dale Crider (pers. comm.) found that wild millet (*Echinochloa*) seeds were important foods in northern Argentina. Eitniear and Rylander (2008) concluded from its bill anatomy that the masked duck is adapted to grazing on relatively tough vegetation, as is typical of many geese. Johnsgard and Carbonell (1996) also suggested that vegetation might be an important food source, judging from the masked duck's rather sturdy bill structure.

Masked ducks dive extremely well and submerge virtually silently, often remaining under water for long periods. They frequently will emerge with only their heads above water and then remain hidden beneath a floating lily pad. They apparently obtain essentially all their foods from vegetable sources, and the ponds that they typically inhabit are not very deep. Almost every pond or marsh on which I have observed masked ducks, or on which masked ducks had been recently seen (in Jamaica, southern Texas, and Colombia), has also had jacanas (*Jacana* spp.) present. This association would suggest the importance of floating-leaf plants as a basic part of the habitat requirements of masked ducks, in conjunction with either the associated foraging opportunities or in providing escape cover.

Least grebes (*Podiceps dominicus*) are often present in masked duck habitats, and the two species are rather similar, both in their diving characteristics and their inconspicuous presence. To what extent the least grebe's foods might overlap with those of the masked duck is unknown, but the least grebe's foods are largely from animal sources. It is of interest that the shape of the masked duck's bill differs so strongly from that of the typical *Oxyura* species and more closely approaches those of some *Anas* and *Aythya* species (Eitniear and Rylander, 2008). One might suppose that this difference might be related to a higher dependence on aquatic plants and a lower incidence of tiny invertebrates, such as midge larvae, in the diet of this species.

Dirk Hagemeyer (in litt.) observed the diving behavior of adults and juveniles. The brood he studied in 1967 always fed in a ditch about three feet deep. They typically foraged for about 15 to 30 minutes and then retired to the grassy shoreline to preen and rest for about 90 minutes. In ten dives that he timed, the young remained submerged from 15 to 17 seconds. The brood he studied in 1969 foraged in the same ditch, where the water was generally 4 to 6 feet in depth and locally up to 8 feet deep. One adult was observed foraging for about 45 minutes in water approximately 5 to 6 feet deep. It remained submerged from 23 to 26 seconds during its foraging dives and had intervening rest periods of 9 to 12 seconds at the surface. In general, the birds preferred to remain about in small areas of open water about 3 to 4 feet deep and scarcely moved more than a few hundred feet over a period of several weeks.

Although some observers have stated that masked ducks can readily "leap" from the water into flight, this behavior is evidently not comparable to the takeoff behavior of dabbling ducks. Dale Crider informed me that he never saw such a leaping takeoff, and that the birds always initially made a shallow dive under water and emerged in flight a foot or so ahead of the point of submergence. This interesting method of taking flight is no doubt related to the posterior location of the feet and the small surface area of the wings, which prevent their effective use in pushing the body out of the water directly. Crider found that in waters too shallow for the birds to dive into, they were unable to take flight directly and could be readily caught.

Sociality, densities, territoriality. No specific information about sociality, densities, and territoriality is available, but the masked duck is probably comparable to the ruddy duck in these regards. In a few locations, flocks of 100 or more have been seen, but these are very rare. The records of dump-nests would suggest that females might sometimes nest in close proximity to others.

Interspecific relationships. Masked ducks are sometimes found on the same ponds as ruddy ducks, but they seem to prefer more densely vegetated bodies of water and are evidently more plant dependent in their foods. Thus they probably compete little if at all with ruddy ducks. The importance of other potential competitors, predators, and so forth still remains relatively unstudied, but the black-headed duck is an obligatory brood parasite and known to parasitize the nests of masked ducks among many other wetland species. The opportunistically brood-parasitic fulvous whistling-duck and rosybill (*Netta peposaca*) are also common breeders in habitats used by masked ducks in northern Argentina.

General activity patterns and movements. Phillips (1926) mentioned that masked ducks apparently forage by day and fly at night, a situation seemingly typical of stiff-tailed ducks generally. Dale Crider (pers. comm.) often observed masked ducks in flight when it was nearly too dark to see.

Social and Sexual Behavior

Flocking behavior. Most observers have noted that masked ducks are rarely seen in large groups, with a maximum of about 140 reported (Eitniear and Morel, 2012). Phillips (1926) indicated that most groups had less than 10 to 20 birds. Anderson (1999) found an average of 6.7 birds per flock among seven Texas flocks that he periodically observed between fall and spring.

Pair-forming behavior. Almost nothing is known of the sexual displays of this species. Courtship reportedly has been seen during April in Texas (Davis, 1966) and during May in Louisiana. Various sounds have been attributed by various observers to the male, including a repeated *kirri-kirro*, a cock pheasant–like response to diverse loud noises, and a dull, almost inaudible *oo-oo-oo* that was probably the same sound described by Hall (see next section). Females are said to utter repeated hissing noises. Dale Crider (pers. comm.) heard clucking sounds uttered by females and said that displaying males apparently produce weak calls, which he was never able to hear.

Copulatory behavior. The copulatory behavior of masked ducks has not been described. One first-person account and associated photographs of an attempted copulation was provided by Rod Hall (pers. comm.) from the Greater Antilles. He observed a female in open water, seemingly treading water. Later a male appeared and slowly approached the female with his head low and neck fairly large. He then expanded his neck to golf-ball size (Fig. 22, top) and uttered a steady *coo-coo-coo* sound similar to that of a rock dove. The female silently faced the male with her bill somewhat raised and her neck stretched upward. The male's tail

Fig. 22. Masked duck, golf-ball display (top) and normal swimming (bottom)

was not raised during the entire period of several minutes, and the female was apparently silent. The female eventually turned and departed (Johnsgard and Carbonell, 1996).

Nesting and brooding behavior. Too few nests have been found in North America to provide any definite information on incubation behavior. Eitniear (2010) reported a brood of masked ducks in Live Oak County, Texas, in October 2007, the most recent known breeding in Texas. The nests and young that have been seen in Texas have all been observed rather late in the year (July to December), and only female-like birds have been found associated with broods. Current evidence suggests that in northern latitudes the male remains in breeding plumage through October and gradually becomes more female-like during November and December. The male's breeding plumage might be retained longer in Costa Rica and Panama, at least by some birds (Johnsgard and Hagemeyer, 1969).

Dale Crider (pers. comm.) noted that in Argentina masked ducks were also fall breeders, and nesting was associated with rising water levels in rice fields. Males remained in the vicinity of the nest until sometime into incubation but were never seen in association with families. Considerable variation in the timing of the postnuptial molt was evident, since some flightless males were found when others were still in full breeding plumage. After the eggs had hatched, the broods were apparently often brought back to the nest site for night roosting. Molting of adults apparently occurred in natural ponds adjacent to the rice fields.

It is possible that cooperative breeding occurs, since ducklings of two different sizes were noted in an Anahuac National Wildlife Refuge brood (Johnsgard and Hagemeyer, 1969), and two females were seen in company with a group of 15 ducklings in Mexico (M. Farmer, cited in Eitniear, 1999).

Postbreeding behavior. Nothing is known of masked duck postbreeding movements. Irregular movements north into the southern United States possibly are the result of periodic tropical storms, and it has been speculated that birds from western Cuba might move to coastal Mexico during the winter period. The brood found at Anahuac National Wildlife Refuge in 1967 was observed for about 45 days before it disappeared, the young presumably having fledged. Masked ducks were not seen again on the refuge until the following summer (Johnsgard and Hagemeyer, 1969).

Ruddy Duck
Oxyura jamaicensis (Gmelin) 1789

Other vernacular names. Butterball, stiff-tail

Range. (*O. j. jamaicensis*) Breeds from central British Columbia to southwestern Mackenzie District, across the Canadian prairies to the Red River valley of Manitoba, with sporadic breeding in southern Ontario and Quebec, and southward through the western and central United States to Baja California, coastal Texas, and locally eastward to the Great Lakes and St. Lawrence River valley. A local resident in the Central Valley of Mexico and the Caribbean (Greater Antilles, Bahamas, Grand Cayman). Winters in North America from British Columbia south along the Pacific coast to Central America. Also winters inland through the western United States south to Mexico and Central America, along the Atlantic coast from Maine to Florida, and along the Gulf coast west to Mexico. Other races breed south to Argentina.

North American subspecies. *Oxyura j. jamaicensis* (Gmelin): North American Ruddy Duck. Breeds in North America, as indicated above. The race *O. j. rubida* Wilson (1814) was erected to distinguish mainland North American birds from those of the West Indies, but it is of questionable validity. Other resident subspecies occur in high-altitude Andean lakes from Colombia (Andean ruddy duck, *O. jamaicensis andina*) disjunctively south along the Andes from Ecuador to Argentina and Chile (Peruvian ruddy duck, *O. jamaicensis ferruginea*). Although some recent research (Livizey, 1997) has associated *jamaicensis* taxonomically with the South American lake duck, that species (*Oxyura vittata*) is instead almost certainly a member of the maccoa–Australian blue-billed duck evolutionary assemblage (Johnsgard, 1968; Johnsgard and Nordeen, 1981; Callighan and McCracken, 2005).

Measurements. *Folded wing:* Delacour (1959): Males 142–154 mm; females 135–145 mm. Kear (2005): Males 144–158 mm (ave. of 62, 152 mm); females 142–153 mm (ave. of 26, 137 mm).

Culmen: Delacour (1959): Males 39–44 mm; females 37–42 mm. Kear (2005): Males 39–44.9 mm (ave. of 62, 42 mm); females 32.9–42.0 mm (ave. of 26, 41 mm).

Weights (mass). Kear (2005): Males 530–700 g (ave. of 37, 588 g); females 450–845 g (ave. of 22, 571 g). Nelson and Martin (1953): 12 males, ave. 1.3 lb. (589 g); 17 females, ave. 1.1 lb. (498 g). Mumford (1954): 10 males, ave. 1.19 lb. (539 g); 6 females, ave. 1.19 lb. (539 g). Jahn and Hunt (1964): 11 males, ave. 1.06 lb. (481 g); adult females, ave. 1.19 lb. (539 g).

Identification

In the hand. Except for the very rare masked duck, ruddy ducks can be easily distinguished from all other North American ducks by their short wings (under 160 mm); long (at least 75 mm) and numerous (18) tail feathers; and their short (under 50 mm), wide, and flattened bill. The ruddy duck differs from the masked duck in lacking any white on the wings and having an outer toe that is as long as or longer than the middle toe, and a bill length (culmen) that is also longer (more than 35 mm). It is the only North American species in which the bill's nail is narrow and small on the upper mandible surface but wide and recurved below the tip.

In the field. Except during fairly late spring and summer, ruddy ducks of both sexes are in a rather brownish and inconspicuous plumage. On the water they appear as small, chunky diving ducks, with short necks and a long tail that is either flat on the water surface or, especially in males, variably cocked above it. The whitish cheeks are the most conspicuous field marks at this time, but as spring progresses the male assumes and a more reddish brown body plumage, a contrasting black crown, and an increasingly bluish bill, a spectral hue that is structurally produced and not pigment-based (Hays and Habermann, 1969). Females are nondescript brown with streaked brown and buff cheeks; their chunky, thick-necked body shape and low swimming profile are probably the best clues to their identity.

Flying ruddy ducks have unusually short, thick necks and distinctively rapid wingbeats; their small wings must beat rapidly to keep the bird aloft. They perhaps have greater difficulty taking flight than any other North American duck, pattering along the water for a considerable distance before attaining flight speed, and the birds rarely attain high altitudes. Neither sex is notably vocal; the female utters a squeaky threat call, and during display males produce a dull thumping percussive noise that terminates in a weak belch-like croak.

Age and Sex Criteria

Sex determination. The wings of males average only slightly larger than those of females and provide no certain sex criteria, but a folded wing measurement of more than 145 mm indicates a probable male. Thus, entirely unmarked white cheeks or brownish red body feathers are the best outward criteria of sex. Males approaching a year of age are likely to have the remarkable intromittent structure typical of *Oxyura*. The adult penis (Fig. 1I) is extremely long (up to almost 8 inches in *O. vittata*), spirally coiled, and spine-covered, a combination of remarkable traits that are apparently unique to *Oxyura* (Wetmore, 1918, 1975; McCracken, 2000; Coker et al., 2002; Brennen et al., 2007). This feature should allow for easy determination of adult sexes by cloacal observation.

Age determination. Juveniles of both sexes evidently retain their juvenal tail feathers until January or February, a surprisingly long time, so notching at the tips of these feathers should provide a useful aging criterion throughout most of the birds' first year (Bent, 1925). Although the narrow juvenal tail feathers of

Oxyura are not conspicuously notched, in young birds the terminal portion of the shaft is often wholly devoid of barbs near the tip. Further, the tertials of immature birds are straight rather than curved and drooping, the greater tertial coverts are somewhat squared rather than rounded at the tips, and the middle coverts are slightly rough and trapezoidal in shape rather than smooth and rounded (Carney, 1964).

One notable feature of ruddy duck molting patterns is that two tail (rectrix) molts annually are frequent if not regular (Jehl and Johnson, 2004). This species' tail molts are quite prolonged, with two generations of rectrices often present simultaneously. The probable advantage of two overlapping tail molts is that the birds never lack a functional tail for underwater maneuvering while foraging. During molting a sequential loss of alternate rectrices is typical, apparently to reduce the disadvantage of having large gaps in the tail's overall surface area. There have also been a few rare cases found of ruddy ducks undergoing two wing molts within a year, the first occurring after breeding and the second in spring (Hohman, 1996).

Distribution and Habitat

Breeding distribution and habitat. This strictly New World species has a North American breeding range that is similar to those of the canvasback and redhead, and, like the ranges of those species, tends to be disruptive and locally declining.

A single definite record of a brood seen in the Tetlin area of eastern Alaska in 1959 then constituted that state's only known breeding, but since then breeding has also been observed at the Minto Lakes and elsewhere. In Canada, the vicinity of Great Slave Lake would appear to be the approximate northern breeding limit of this species, which increases in abundance southwardly through British Columbia, and southeastwardly through Alberta, Saskatchewan, and Manitoba. There is also local breeding in southeastern Ontario and southern Quebec, especially along the St. Lawrence River valley.

In Washington, breeding by ruddy ducks has long been fairly regular throughout the eastern part of that state (Yocom, 1951) and they have also been a long-time local breeder in the Potholes region of Grant County, central Washington (Johnsgard, 1956). The birds also breed in shallow lakes and marshes of eastern Oregon, especially on the vast wetlands of Harney, Lake, and Klamath Counties, where roughly 12,000 nesting birds have bred recently (Gilligan et al., 1994). In California they have nested locally as far south as the Salton Sea. They also breed in the freshwater marshes of Baja California, and on the arid central uplands of Mexico, south to the Valley of Mexico (Leopold, 1959). Limited breeding occurs along the southern coast of Texas and perhaps also in interior Texas. There are also scattered records of nesting in Kansas, New Mexico, Arizona, and Nevada.

The heart of the ruddy duck's breeding range is in the prairie marshes of the northern Great Plains, extending westward from Minnesota (Lee et al., 1964a) and northwestern Iowa (Low, 1941), north through Nebraska, both Dakotas, and eastern Montana to Manitoba, Saskatchewan, and eastern Alberta (Salt and Salt, 1976; Semenchuk, 1992; Smith, 1996).

Along the Gulf and Atlantic coasts breeding records exist for Louisiana, Florida, the Carolinas, Maryland, Delaware, and New Jersey. Local breeding has also been reported in Indiana, Ohio, and Pennsylvania.

The breeding (hatched, with denser concentrations inked), wintering (shaded), and marginal (stippled) range of the ruddy duck.

In New York ruddy ducks bred at Jamaica Bay from the 1950s to the mid-1980s, as well as in the Montezuma National Wildlife Refuge marshes. By the 1960s they had bred in Indiana, Wisconsin, and Michigan, where nesting has more recently been reported from more than a dozen counties (Brewster, McPeek, and Adams, 1991).

The breeding habitat of ruddy ducks consists of permanent freshwater or alkaline marshes having emergent vegetation and relatively stable water levels. Suitable nesting habitat must have open water in fairly close proximity to nesting cover, including emergent plants that provide accessibility as well as adequate cover density. These plants additionally can be bent down by the birds to form a nest platform and allow water passageways, such as muskrat channels, that permit easy movements between the nest and open water (Bennett, 1938; Joyner, 1969).

Population. Published population estimates of ruddy ducks are limited, but unpublished data from the US Fish and Wildlife Service and the Canadian Wildlife Service from 1955 to 2000 suggest that the prairie pothole region of the Great Plains then supported a breeding population of about 410,000 birds. Extreme yearly estimates ranged from 170,000 to 950,000 birds and suggest an apparently slightly upward long-term population trend (Baldassarre, 2014).

An average of about 53,000 birds were shot annually during the 2013 and 2014 US hunting seasons (Raftovich, Chandler, and Wilkins, 2015). In the United States the long-term average annual kill averaged about 50,000 birds until about 2000. The average annual kill in Canada from 1974 to 1999 approximated 3,700 birds (Brua, 2002).

Ruddy ducks were introduced into English bird collections in the 1940s, and by 1961 some had escaped into the wild. They later spread to Spain, and still later elsewhere in western Europe and northern Africa. In 2005 an eradication program began in the United Kingdom, and later also began in France, Spain, and Portugal (Henderson, 2010a). However, as of 2014 ruddy ducks have been reported from at least nine European and North African countries during the breeding season, and have been sighted in at least 13 more. Eradication programs in the United Kingdom have proved very effective, but continental European populations are probably still increasing, putting the endangered white-headed duck (*Oxyura leucocephala*) at substantial genetic risk through hybridization with ruddy ducks. These hybrids are known to be fertile.

Wintering distribution and habitat. Wintering in Canada is limited to small numbers in southern British Columbia and southern Ontario (Godfrey, 1986). Winter surveys in the United States and Mexico during the late 1960s indicated that more than 60 percent of the wintering ruddy duck population occurred in the Pacific Flyway, with the Mississippi and Atlantic Flyways each providing about 15 percent, and the Central Flyway accounting for less than 10 percent. Midwinter surveys from 2000 to 2010 indicated that 50 percent of the 190,000 ruddy ducks then wintered in the Pacific Flyway, with 40 percent in the Atlantic Flyway, and the remaining 10 percent equally divided between the Mississippi and Atlantic Flyways (Baldassarre, 2014).

In the Pacific Flyway, ruddy ducks winter in Puget Sound and southward along the coastline of

Washington, Oregon, and especially California, where more than 80 percent were wintering in the early 2000s. They are largely confined to brackish bays and freshwater areas, and are rarely seen on the open ocean.

Ruddy ducks were historically abundant on the Pacific coast of Mexico during winter, where among diving ducks they were once outnumbered only by the lesser scaup. Leopold (1959) mentioned seeing a flock of more than 107,000 birds in a single lagoon near Acapulco in 1952 and stated that they thrived in the brackish coastal marshes of Mexico. Relatively few then occurred on Mexico's Gulf coast during winter, and even fewer were found in the interior of Mexico. Counts in Mexico since that time suggest far smaller populations; surveys made in the early 2000s found only about 6,000 birds from the Baja Peninsula south to Nayarit, about 12,000 in the internal highlands, and about 14,000 on the eastern coast of Mexico (Baldassarre, 2014).

On the Atlantic coast, ruddy ducks winter from Maine southward, occurring as far south as Florida's Lake Okeechobee and on the Kissimmee River valley lakes, as well as on brackish coastal marshes of the Gulf coast from Florida through Louisiana and Texas. Ruddy ducks also winter in small numbers throughout the interior of these states but not in the numbers typical of coastal situations. Midwinter Waterfowl Survey counts in the Atlantic Flyway from 2000 to 2010 totaled nearly 190,000 birds, with almost half of them in the Chesapeake Bay region plus 30 percent in Virginia, 11 percent in North Carolina, and 5 percent in Florida (Baldassarre, 2014).

In the Chesapeake Bay region, where the average wintering numbers of ruddy ducks have long composed 40 to 50 percent of the total Atlantic Flyway winter population, their ecological distribution is of interest. January counts during the 1950s indicated that slightly brackish estuarine bays supported 54 percent of the birds, brackish estuarine bays 41 percent, salt estuarine bays 5 percent, and freshwater estuarine bays 1 percent. Ruddy ducks wintering in the Chesapeake Bay region avoid the coastal bays and ocean proper, and apparently move to saltwater estuarine bays only during the coldest weather (Stewart, 1962). It thus appears that ideal ruddy duck wintering habitat on the Atlantic coast consists of brackish to slightly brackish estuaries or coastal lagoons of shallow depths. An abundance of submerged plants, small mollusks, and crustaceans no doubt also figures importantly in winter usage by ruddy ducks.

During the 2000–10 Mississippi Flyway Midwinter Waterfowl Surveys, more than two-thirds of the nearly 10,000 wintering ruddy ducks were concentrated in Mississippi, especially on Arkabutla Lake and at Yazoo National Wildlife Refuge, and a fifth were observed in Tennessee, mostly at Reelfoot Lake (Baldassarre, 2014).

General Biology

Age at maturity. McClure (1967) stated that captive-raised male ruddy ducks mature their first year, and although two females also bred their first year, captive females usually do not breed until their second year. Ferguson (1966) noted that two aviculturists reported breeding by first-year birds, two in the second year, and one in the third year. It is likely that first-year breeding is frequent if not regular in wild ruddy ducks. However, Siegfried (1976a) reported that 49 percent of 35 one-year-old females did not nest in his Manitoba study, whereas 85 percent of 61 older birds did.

Fig. 23. Ruddy duck, adult male displaying to female

Pair-bond pattern. Pair-bonds in ruddy ducks are weak or lacking; the males seem far more prone to fight and defend territories than to form strong pair-bonds (Gray, 1980; Carbonell, 1983; Johnsgard and Carbonell, 1996). Polygyny or promiscuity seems to be the usual stifftail mating system. In at least some *Oxyura* species (*O. maccoa, O. australis, O. leucocephala, and O australis*) as well as in the musk duck (*Biziura lobata*) a marked degree of sexual dimorphism exists, with an average male-to-female adult mass ratio of about 1.6:1 (Kear, 2005).

The male-favored sexual dimorphism and unique behavioral traits of all these species can largely be explained by the intense sexual selection brought about by a promiscuous mating system, with exaggerated male size and intense male-to-male aggression, rather than highly developed intersexual attraction behavior, strongly influencing male appearance and behavior (Johnsgard, 1966; Johnsgard and Carbonell, 1996).

Nest location. Williams and Marshall (1938) observed that among 50 ruddy duck nests in Utah, 32 percent were in hardstem bulrush (*Scirpus acutus*), an amount well in excess of the plant's relative abundance. About the same number (30 percent) were in alkali bulrush (*S. paludosus*), a much more common species of bulrush. About 20 percent were in salt grass (*Distichlis*), a surprisingly high percentage considering that this is not a true emergent species and would provide little if any overhead cover.

Bennett (1938) found that 14 of 22 Iowa nests were in stands of roundstem bulrush (*S. occidentalis*), 6 were in mixed stands of this species and other emergent vegetation, 1 was among reeds (*Phragmites*), and 1 was in a mixture of emergent plants and sedges. Bennett believed that roundstem bulrush was favored for nesting because of the relative ease with which it can be bent over to form a nest. Low noted, as had Bennett, that river bulrush (*Scirpus fluviatilis*) was not used for nesting, probably because its stiff stalks make nest building difficult. Joyner (1969) noted that three of nine nests were in Olney bulrush (*S. olneyi*) or mixed bulrush-cattail stands, one each was in hardstem and alkali bulrush stands, one in a mixture of cattails and hardstem bulrushes, one in cattails, and two in salt grass.

Low (1941) examined 71 Iowa nests and concluded that nesting cover was determined not so much by preferences for specific plants as for cover types having a suitable water depth. A depth of 10 to 12 inches at the nest location was favored, with an observed range of 0 to 36 inches. Cover density at the nest site was dense in 63 percent of the nest sites and sparse in only 6 percent. Based on the total numbers of nests, lake sedge (*Carex lacustris*) provided cover for the most, followed by hardstem bulrush and narrow-leaved cattail (*T. angustifolia*). On the basis of usage relative to available cover, the relative plant cover choice in decreasing sequence was slender bulrush (*S. heterochaetus*), whitetop (*Fluminea*), hardstem bulrush, lake sedge, and narrow-leaved cattail.

Clutch size. Low (1941) found an average clutch size of 8.1 eggs in 71 nests; Bennett (1938) noted an average clutch of 7.05 in 18 nests; and Williams and Marshall (1938) found 158 eggs in 19 nests, or 8.3 eggs per nest. Frequent parasitic egg laying (including apparently random "dump-nesting") influences average clutch size data and helps account for some unusually large clutches. Joyner considered that any nest containing more than 10 ruddy duck eggs had been affected by intraspecific parasitism. The egg-laying rate is probably about one per day, with a day sometimes being missed (Joyner, 1969; Siegfried, 1976a; Gray, 1980).

Incubation period. Ruddy duck eggs hatched in an incubator required 21 to 25 days to hatch (Hochbaum, 1944). Low (1941) determined that four naturally incubated clutches hatched in 25 days and two in 26 days. Joyner (1969) estimated the incubation period of eggs incubated by wild birds to be 23 to 24 days.

Fledging period. Hochbaum (1944) considered that 52 to 66 days are probably required for fledging. Siegfried (1973b) reported that captive-raised birds fledged at 7 to 8 weeks, and Helen Hays similarly judged that fledging by wild birds required 6 to 7 weeks (Bellrose, 1980).

Nest and egg losses. Low (1941) determined that water-level fluctuations were the most serious source of nest and egg losses in Iowa, with rising levels causing nest flooding and declines causing nest desertion. Of 71 nests he observed, 52 (73 percent) were successfully hatched, 12 were deserted, 4 were flooded, and 3 were lost to mink predation. Williams and Marshall (1938) observed a lower nesting success (38 percent) in Utah, with no losses caused by predation but 48 percent of the eggs lost to other factors, such as desertion and flooding.

Fig. 24. Ruddy duck bubbling display sequence (A–I), followed by ringing rush display (K–Q), and precopulatory bill-flicking (R–S).

An example of nesting success involving larger sample sizes includes those of Brua (1999), who reported a 41 percent nesting success among 233 Minnesota nests but also noted (2002) that, considering a variety of studies, the most frequent causes of nest loss were predation, flooding, and desertion. Brua concluded that nesting success for his study area was unpredictable, probably because the diversity of local predators precluded a choice of safe nest sites.

Hatching success rates (percent of eggs hatching) tend to be fairly high in ruddy ducks. As summarized by Brua (2002), in various studies they have included 69 percent success for 379 Iowa eggs, 71 percent for 472 Utah eggs, 74 percent for 268 California eggs, and 98 percent for 125 eggs in another Iowa sample. Embryonic death is a frequent cause of egg failure, perhaps as a result of incubation interruptions caused by brood parasitism. Renesting is apparently rare in ruddy ducks, but Tome (1987) reported one case of a renesting female that had lost her first clutch of 11 eggs and replaced it with a clutch of 9 eggs after a renesting interval of about four days.

Juvenile mortality. Because of seemingly weak parental attachment, ruddy duck broods rarely retain their original composition for very long, and thus brood size counts fail to provide a suitable estimate of pre-fledging mortality. Joyner (1969) noted a tendency for abandoned ducklings to join with other ducklings, especially those somewhat older than themselves, and believed that because of their high degree of precocity they seem to survive well when separated from adults.

Male ruddy ducks remain sexually attracted to females for a relatively longer period than do most North American dabbling ducks, often displaying to females attending broods. This prolonged sexual attraction has probably led people to believe that second broods might be raised during a single summer in southern regions (Kortright, 1943), but the only North American waterfowl species for which double-brooding has been proven is the black-bellied whistling-duck.

Estimates of juvenile survival include those of Brua (1998), who determined an average brood survival rate of 84.3 percent for 57 broods over a three-year study period. During two different years, Pelayo (2001) found the duckling survival rates over the first 30 days of life during two breeding seasons to be 49 percent (76 birds) and 76 percent (168 birds). These relatively high survival rates might be related in part to the notably high level of precocity that is typical of *Oxyura* ducklings. They are able to dive well when only a few days old, and might become independent long before fledging, often by about three weeks of age. Predators in Utah included ring-billed gulls, California gulls, and black-crowned night-herons (Joyner, 1977b).

Postfledging mortality rates of juveniles are not yet available, but a highly disproportionate age ratio among 387 birds shot in Wisconsin (5.2 immatures per adult) suggests a much higher vulnerability to hunters among immatures than in adult birds (Jahn and Hunt, 1964).

Adult mortality. No estimates of annual adult mortality rates are available, probably because it is so difficult to capture and band large numbers of ruddy ducks. Banded individuals (possibly from the introduced European population, which is considered a pest and invasive species) have been known to survive up to 13 years in the wild (Kear 2005); in North America a wild male similarly survived 13 years and 7 months

Ruddy duck, breeding male swimming

(Baldassarre, 2014). In a 2007 summary, only 550 ruddy duck bands had by then been recovered and reported to the US Fish and Wildlife Service's Bird Banding Laboratory. By comparison, more than 1 million mallard bands have been recovered in North America, out of a total of about 7 million mallards banded, a remarkably high recovery rate.

General Ecology

Food and foraging. Cottam's (1939) study of foods found in 163 adult ruddy ducks taken over nine months of the year is the most comprehensive analysis to date. He found plant foods to constitute more than 70 percent by volume of the materials found, with pondweeds (Najadaceae) and sedges (Cyperaceae) accounting for nearly half of the total. Tubers, stems, and leaves of pondweeds, especially *Potamogeton* species, seem to be the favorite foods, whereas bulrushes (*Scirpus*) figured most prominently among the sedges that were represented.

A large number and variety of insects are also consumed, especially during summer. Of these, the larvae of midges (Chironomidae) are particularly important, especially during summer (Cottam, 1939; Woodin and Swanson, 1989), perhaps because of their abundance in mud-bottom waters. Lynch (1968) likewise mentioned that midge larvae are consumed in the Chesapeake Bay region. Other aquatic and terrestrial insects are also sometimes eaten, even including a few land-dwelling forms such as locusts. Cottam found relatively few mollusks and similarly small amounts of crustaceans among the samples he examined. It is probable that large samples from brackish-water wintering areas would have greater quantities of these organisms present.

In that regard, Stewart (1962) provided information on food contents of 35 ruddy ducks from the Chesapeake Bay region. He noted that the seeds, leaves, and stems of various submerged plants, and certain small mollusks and crustaceans, were the principal foods present in these samples. Small bivalve macoma mollusks (*Macoma* spp.), small *Mya* and *Mulinia* clams, the gastropod *Acteocina*, and various amphipod and ostracod crustaceans were represented frequently in these samples.

Water depths favored and normal diving times have been little studied in ruddy ducks. A few observations on captive ruddy ducks (Johnsgard, 1967) indicated average diving times of about 14 seconds, with intervening pauses of 10 seconds, and a maximum observed dive duration of 29 seconds These data were from birds foraging in the shallow and turbid ponds at England's Wildfowl Trust (now Wildfowl and Wetlands Trust). Siegfried (1973) observed that wild ruddy ducks similarly foraged in water up to one meter deep and their dives averaged 18.6 seconds (males) and 20.6 seconds (females). The strongly flattened bill of this species seems well developed for probing in muddy bottoms and sifting out small particles, and its slightly recurved nail might be useful in tearing leaves or stems from underwater plants.

Sociality, densities, territoriality. Low (1941) estimated that the average nesting density was about 1 nest per 21 acres over 1,000 acres of nesting cover. In some areas the nest densities approached 1 nest per 10 acres, and the maximum observed density was 1 nest per 2.4 acres on a 32-acre Iowa marsh. Williams and Marshall (1938) estimated an overall nesting density of 1.6 nests per 100 acres over a total of 3,000 acres of potential nesting cover in Utah, with the highest observed density being 2 nests on 1.5 acres of hardstem bulrushes. In reviewing breeding densities of waterfowl on five prairie areas in Canada and South Dakota, Stoudt (1969) reported a range of broad-scale ruddy duck densities of from less than 1 to as many as 12 pairs per square mile.

To a greater degree than is characteristic of other North American diving ducks, the ruddy duck appears to occupy and defend a definable territory during the breeding season, or at least a definable area surrounding the nest site (Joyner, 1969; Siegfried, 1976a). The performance of the "bubbling" display by breeding males in the absence of a female, and often when other males come into view, leads one to believe that this display serves also as a territorial pronouncement, rather than simply as a heterosexual attraction signal. Joyner (1969) believed that the male defends a small territory that might extend only as far as about 10 feet around the nest. He observed that these apparent territories were spaced from 20 to 100 feet apart and that territorial dispersion was influenced by the presence or absence of channels through the emergent vegetation.

Ruddy duck, breeding male and duckling

Interspecific relationships. Evidently the presence of foreign eggs in ruddy duck nests affects nest desertion rates and thus nesting success. Reinecker and Anderson (1960) found that 33 percent of ruddy duck nests that had been abandoned contained redhead eggs, but among successfully hatched ruddy duck nests only 11 percent contained redhead eggs. Low (1941) noted that 11 (12.6 percent) of the ruddy duck nests he found had 1 to 4 redhead eggs present, none of which hatched. Joyner (1969, 1976) observed mixed clutches that included eggs of other ruddy ducks, cinnamon teal, mallards, and northern pintails, although it was in some cases impossible to determine which species initiated the nest. He found that the hatching success of parasitically laid ruddy duck eggs among nests of four host duck species ranged from 18 to 67 percent, but only 10 percent of the redhead eggs laid in ruddy duck nests hatched.

Using molecular techniques to establish female parentage, Reichert et al. (2010) found that 56 of the 112 females in the population laid some parasitic eggs, although most of the females laying parasitic eggs also deposited eggs in their own nests, sometimes at different stages of their own clutch completion. Furthermore, 67 percent of all the ruddy duck nests they studied contained mixed-parentage eggs.

Ruddy duck eggs have been found in nests of the gadwall, northern pintail, cinnamon teal, canvasback, redhead, and ring-necked duck, as well as in nests of the American coot and pied-billed grebe. Eggs of cinnamon teal, canvasbacks, and redheads have likewise been found in ruddy duck nests (Brua, 2002).

Joyner (1969, 1973, 1975, 1983) estimated that brood parasitism rates by ruddy ducks ranged from 2 to 11 percent of mallard, northern pintail, cinnamon teal, and redhead nests in Utah. He reported that ruddy ducks collectively parasitized 7.7 percent of 809 nests of these four duck species; parasitically laid ruddy duck eggs composed 9 percent of all 1,449 ruddy duck eggs that he found. The hatching success of all ruddy duck eggs incubated by ruddy females (including any eggs that might have resulted from conspecific parasitism) was 31 percent, whereas 10 percent of ruddy duck eggs deposited in 42 cinnamon teal nests hatched, as did 18 percent in 10 redhead nests, 36 percent in 5 mallard nests, and 92 percent in 4 northern pintail nests. These figures indicate a collective average hatching success of less than 20 percent for all the ruddy duck eggs that were incubated by other species.

In general, waterfowl are relatively inefficient brood parasites, as compared with most species of obligatory brood parasites. A more extended discussion of avian brood parasitism can be found elsewhere (Johnsgard, 1997), including the breeding biology of the South American black-headed duck (*Heteronetta atricapilla*), the only obligate brood parasite species among waterfowl.

General activity patterns and movements. Foraging by male stifftails at the Wildfowl Trust occupied 13.6 to 27.0 percent of their daytime activities. Sexual display occupied another 5.5 to 14.5 percent (7.1 percent in the case of male ruddy ducks), and was most frequent after 4:00 p.m. Another 12.2–19.2 percent was spent in comfort activities, and the rest of the daytime hours were spent in swimming and resting. During one night's observations of ruddy ducks, male display behavior continued until about 10:00 p.m., when foraging began (Johnsgard and Carbonell, 1996). Since midge larvae tend to emerge from the pond's substrate and move into the water column at night, such nocturnal foraging is adaptive in their exploiting of relative food availability.

Flocking behavior. Although it rarely associates closely with other species, the ruddy duck sometimes gathers into large flocks during migration and in favored wintering areas. An immense flock of more than 107,000 wintering birds seen by Leopold (1959) in Mexico was noted earlier, and Stewart (1962) reported counts of several thousand birds during winter and spring in the Chesapeake Bay region.

Pair-forming behavior. In association with the relatively late assumption of the nuptial plumage by males, a relatively late period of pair formation is typical. I (1955) detected no obvious pair-bonding among wild birds in central Washington, and interpreted the male displays I observed there in late May and June as territorial advertisement or sexual-readiness signaling. Joyner (1969) observed male displaying from April to late July and interpreted much of the later displays as representing territorial advertisement or defense. He noted that this species normally began courtship activities only after arrival on the breeding grounds rather than while wintering or migrating.

Ruddy duck, male tail-cocking display

The most frequent of the male's sexual-signaling and/or territorial-proclamation displays is the "bubbling" display (Johnsgard, 1965; Johnsgard and Carbonell, 1996). This display (Fig. 24A–I) is usually described as a rapid, repeated beating of the bill on the highly expanded neck, and it depends on the presence of an air sac that is inflatable via a connection to the trachea (Wetmore, 1919). Although all the other species of *Oxyura* can expand their neck profile to varying degrees (most obviously the masked duck), it is not known how many other *Oxyura* might have tracheal air sacs; esophageal inflation would produce the same visual effect, as would neck feather erection.

The ruddy duck's bubbling display has unique acoustical and visual characteristics that make it well adapted for a mostly vegetated marshy habitat, including both the percussive tapping sounds it produces and the ring of bubbles formed in front of its breast feathers as air is forced out from them. A belch-like sound terminates the bubbling display as the head is brought forward and the bill is opened (Fig. 24I), perhaps as a result of air being released from the tracheal sac.

Although bubbling is most often performed before females or (rarely) toward other males, males also perform it in the apparent absence of other birds, suggesting a territorial advertisement function. Gray (1980) noted that males performed bubbling mostly in social rather than nonsocial situations, most often prior to the start of the nesting season, and most frequently directed toward females.

Males also often try to swim directly in front of females while cocking their tails vertically and exposing their white under tail-coverts ("tail-flashing" display), often stopping momentarily to perform the bubbling display. Courted females performed no obvious inciting behavior but instead responded aggressively. A soft squeaking noise often accompanied their gaping threats, and was almost the only evident sound made by courted female ruddy ducks. Males sometimes also utter very soft aggressive notes.

Other less frequent behaviors of males during social encounters include an alternated bill-dipping, and head-flicking sequence (Fig. 24R–S) that is mostly used in precopulatory situations (see the next section), head-dipping, wing-shaking, head-shaking, and cheek-rolling. All of these actions differ little, if at all, from their functional counterparts as comfort or maintenance activities and are of generally uncertain significance as possible social signals.

Displaying ruddy ducks also often perform quick aggressive rushes over the water toward other nearby males, followed by equally rapid returns to the female. A short display flight, or "ringing rush," along the water surface toward the female is also frequently performed (Fig. 24K–Q) and produces even more water-splashing noises ("ringing") than does the bubbling display.

It is of interest that Peruvian ruddy ducks (*O. jamaicensis ferruginea*) I have observed in various zoos also performed essentially the same sexual displays as do male North American ruddy ducks, including an abbreviated bubbling sequence, with an average of about three tapping movements per sequence rather than seven. The Colombian race has not been analyzed in this regard, but first-generation hybrids between the Peruvian and North American races were closer to the American race in the number of taps performed. Male Peruvian ruddy ducks likewise exhibit none of the distinctive male postural displays typical of the outwardly similar Argentine lake duck (*Oxyura vittata*), maccoa duck (*O. maccoa*), and Australian blue-billed duck (*O. australis*), suggesting that their shared black head plumage coloration is coincidental and does not indicate close phyletic affinities with *ferruginea* (Johnsgard, 1965, 1966, 1968). Hughes (2005) likewise included *ferruginea* in *O. jamaicensis*, whereas Livizey (1995) regarded it as a separate species.

Copulatory behavior. M. Carbonell and I have observed several copulation sequences (Johnsgard, 1965; Carbonell, 1983; Johnsgard and Carbonell, 1996). The male typically approached the female cautiously, periodically dipping his bill and flicking it laterally when it was most retracted (Fig. 24R–S). Sometimes the female would assume a partially prone posture, but in most of the cases the male suddenly mounted her without any obvious indication of readiness on the part of the female. During treading the female was almost entirely submerged, and as soon as copulation was terminated the male dismounted, faced the female, and performed the bubbling display several times in quick succession. A lengthy preening period followed.

In one case, I observed that the male's amazingly long phallus (see Fig. 1 and Coker et al., 2002) remained fully extended for several minutes after copulation had terminated. Brennen et al. (2007) reported that the

Ruddy duck, male bubbling display

extended phallus length of *O. jamaicensis* is 12 to 18 cm and suggested that waterfowl phallus lengths are positively correlated with the frequency of forced extra-pair copulations.

It has also been speculated that the complex coiled configuration and surface attributes of the phallus among male waterfowl might function by removing from the female's reproductive tract any sperm still present from previous matings (Briskie and Montgomerie, 1997). Judging from personal observations, rape behavior is uncommon in captive ruddy ducks, although Gray (1980) observed numerous examples involving wild birds. Among 25 females subjected to attempted rape, 84 percent were unpaired, and 25 percent of the rape attempts were successful. Both paired and unpaired males were involved in these attempts, and they involved both paired and unpaired females (Gray, 1980).

Nesting and brooding behavior. Egg-laying begins as soon as a flimsy nest foundation has been laid. As the clutch increases in size, the nest is gradually enlarged, and often both an entrance "ramp" and an overhead cupola are added by manipulating the surrounding vegetation. There is usually little or no down present, and the nest might either have a well-developed bowl or be nearly flat on the nesting platform. Evidently incubation might begin before the last egg is deposited because it is not uncommon for eggs with incomplete embryonic development to be left at the time the female departs with her newly hatched brood. The incubation rhythm of the female is still little known, but the occurrence of only slightly incubated eggs among hatched clutches suggests that the female leaves her nest for sufficiently long periods as to allow other females to surreptitiously add eggs to the host female's clutch.

Joyner (1969) observed that ruddy duck broods in Utah hatched from mid-May to early August, suggesting that time might be limited for successful renesting if the initial nesting was unsuccessful. Bellrose (1980) provided early evidence of renesting by ruddy ducks, noting that a Manitoba female renested within seven days after losing her first clutch. Renesting was later also documented by Tome (1987), who observed a case of renesting in Manitoba, but in general renesting efforts seem to be very rare among ruddy ducks. Likewise, there is no evidence to support a long-held supposition (Kortright, 1942) that two broods might be reared by ruddy ducks in southern parts of their US range.

There is also no indication that male ruddy ducks seen with broods protect them, or are paternally related to them (Gray, 1980). Joyner (1969) saw drakes regularly following families until late June. Only one of 22 broods he counted on July 10 was associated with a male adult, although 15 of these broods were then still being led by females. I have often watched breeding-plumage males following females with broods but have never observed any helping or protective behavior on the male's part. Rather, I observed only sexual displays by the male and responding hostile behavior by the female; one duckling that started to follow the male was avoided by it.

Postbreeding behavior. Mothers typically abandon their broods at unusually early ages, averaging only 20.1 days in one study (Brua, 1998), but young stifftails are highly precocial, and by three weeks of age they probably can survive independently. Most females left Joyner's Utah study area by late July and apparently moved to molting areas, although a few broods were still hatching at that time.

Considering the limited flying abilities of ruddy ducks, it seems probable that adults do not move far from their breeding areas to undergo their molts. Yet there are few observations on ruddy ducks during this period. They probably are extremely secretive, inhabiting the densest cover in overgrown marshes. They no doubt normally undergo their flightless periods at this time, but scattered reports of flightless ruddy ducks during winter and spring months suggest that an unusual wing molt pattern might be present. A double annual wing molt has been noted by Siegfried (1971) in at least one other *Oxyura* species, but it evidently occurs only rarely in ruddy ducks (Hohman, 1996; Hobson et al., 2000).

III. References

The Birds of North America Monographs

(See also online versions available at the Cornell Laboratory of Ornithology website, http://bna.birds.cornell.edu/)

Austin, J. E., C. M. Custer, and A. D. Afton. 1998. Lesser scaup (*Aythya affinis*). *The Birds of North America* 338. (A. Poole and F. Gill, eds.). The Academy of Natural Sciences, Philadelphia, PA, and American Ornithologists' Union, Washington, DC.

Brua, R. B. 2002. Ruddy duck (*Oxyura jamaicensis*). *The Birds of North America* 696. (A. Poole and F. Gill, eds.). The Academy of Natural Sciences, Philadelphia, PA, and American Ornithologists' Union, Washington, DC.

Eitniear, J. C. 1999. Masked duck (*Nomonyx dominica*). *The Birds of North America* 393. (A. Poole and F. Gill, eds.). The Academy of Natural Sciences, Philadelphia, PA, and American Ornithologists' Union, Washington, DC.

Hohman, W. L., and R. T. Eberhardt. 1998. Ring-necked duck (*Aythya collaris*). *The Birds of North America* 329. (A. Poole and F. Gill, eds.). The Academy of Natural Sciences, Philadelphia, PA, and American Ornithologists' Union, Washington, DC.

Hohman, W. L., and S. A. Lee. 2001. Fulvous whistling-duck (*Dendrocygna bicolor*). *The Birds of North America* 562. (A. Poole and F. Gill, eds.). The Academy of Natural Sciences, Philadelphia, PA, and American Ornithologists' Union, Washington, DC.

James, J. D., and J. E. Thompson. 2001. Black-bellied whistling-duck (*Dendrocygna autumnalis*). *The Birds of North America* 578. (A. Poole and F. Gill, eds.). The Academy of Natural Sciences, Philadelphia, PA, and American Ornithologists' Union, Washington, DC.

Kessel, B., D. A. Rocque, and J. S. Barclay. 2002. Greater scaup (*Aythya marila*). *The Birds of North America* 650. (A. Poole and F. Gill, eds.). The Academy of Natural Sciences, Philadelphia, PA, and American Ornithologists' Union, Washington, DC.

Mowbray, T. B. 2002. Canvasback (*Aythya valisineria*). *The Birds of North America* 659. (A. Poole and F. Gill, eds.). The Academy of Natural Sciences, Philadelphia, PA, and American Ornithologists' Union, Washington, DC.

Woodin, M. C., and T. C. Michot. 2002. Redhead (*Aythya americana*). *The Birds of North America* 695. (A. Poole and F. Gill, eds.). The Academy of Natural Sciences, Philadelphia, PA, and American Ornithologists' Union, Washington, DC.

Taxonomic Works and General Studies

American Ornithologists' Union. 1957. *Check-list of North American Birds.* 5th ed. American Ornithologists' Union, Washington, DC.

American Ornithologists' Union. 1998. *Check-list of North American Birds.* 7th ed. American Ornithologists' Union, Washington, DC.

Baldassarre, G. A, 2014. *Ducks, Geese and Swans of North America.* Rev. ed. Johns Hopkins University Press, Baltimore, MD.

Baldassarre, G. A., and E. G. Bolen. 2006. *Waterfowl Ecology and Management.* 2nd ed. Krieger Publishing, Malabar, FL.

Bannerman, D. A. 1958. *Birds of the British Isles*, vol. 7. Oliver & Boyd, Edinburgh and London, UK.

Bauer, K. M., and U. N. Glutz von Blotztheim. 1968–1969. *Handbuch der Vogel Mitteleuropas.* Vols. 2 and 3. Akademische Verlagsgesellschaft, Frankfurt am Main, Germany.

Bellrose, F. 1976. *The Ducks, Geese and Swans of North America.* 2nd ed. Wildlife Management Institute, Washington, DC.

Bellrose, F. 1980. *The Ducks, Geese and Swans of North America.* 3rd ed. Wildlife Management Institute, Washington, DC.

Bent, A. C. 1925. *Life Histories of North American Wild Fowl.* Part 2. US National Museum Bulletin 130. US Government Printing Office, Washington, DC.

Bezzel, E. 1959. Beiträge zur Biologie der Geschlecter bei Entenvogeln. *Anzeiger der Ornithologischen Gesellschaft im Bayern* 5: 269–355.

BirdLife International. 2000. *Threatened Birds of the World.* Lynx Editions, Barcelona, and BirdLife International, Cambridge, UK.

Bolen, E. G., and M. K. Rylander. 1983. *Whistling-Ducks: Zoogeography, Ecology, Anatomy*. Special Publication 20, The Museum, Texas Tech University. Texas Tech Press, Lubbock.

Bond, J. 1971. *Birds of the West Indies*. 2nd ed. Houghton Mifflin, Boston, MA.

Brennan, L. R., R. O. Prum, K. G. McCracken, M. D. Sorenson, R. E. Wilson, and T. R. Birkhead. 2007. Coevolution of male and female genital morphology in waterfowl. *PLOS ONE* 2(5): 418.

Brennan, P. L. R., C. J. Clark, and R. O. Prum. 2010. Explosive eversion and functional morphology of the duck penis supports sexual conflict in waterfowl genitalia. *Proceedings of the Royal Society B: Biological Sciences* 277(1686): 1309–1314.

Briskie, J. V., and R. Montgomerie. 1997. Sexual selection and the intromittent organ of birds. *Journal of Avian Biology* 28:73–86.

Canadian Wildlife Service Waterfowl Committee. 2013. *Population Status of Migratory Game Birds in Canada*. Canadian Wildlife Service Migratory Birds Regulatory Report No. 40. Canadian Wildlife Service, Ottawa, ON. http://www.ec.gc.ca/rcom-mbhr/default.asp?lang=En&n=B2A654BC-1

Carney, S. M. 1964. *Preliminary Key to Age and Sex Identification by Means of Wing Plumage*. US Fish and Wildlife Service Special Scientific Report: Wildlife No. 82.

Coker, C. R., F. McKinney, H. Hays, S. V. Briggs, and K. M. Cheng. 2002. Intromittent organ morphology and testis size in relation to mating systems in waterfowl. *Auk* 119: 403–413.

Cramp, S., and K. E. L. Simmons, eds. 1977. *Handbook of the Birds of Europe, the Middle East, and North Africa: The Birds of the Western Palearctic, Volume 1, Ostrich to Ducks*. Oxford University Press, Oxford, UK.

Delacour, J. 1954–1964. *The Waterfowl of the World*. 4 vols. Country Life, London, UK.

Delacour, J., and E. Mayr. 1945. The family Anatidae. *Wilson Bulletin* 57: 3–55.

del Hoyo, J., A. Elliott, and J. Sargatal. 1992. *Handbook of the Birds of the World. Vol. 1. Ostrich to Ducks*. Lynx Edicions, Barcelona, Spain.

Dementiev, G. P., and N. A. Gladkov, eds. 1967. *Birds of the Soviet Union*. Israel Program for Science Translations, Jerusalem [translated from Russian].

Gillham, E., and B. Gillham. 1996. *Hybrid Ducks: A Contribution towards an Inventory*. Hythe Printers, Hythe, Kent, UK.

Godfrey, W. E. 1986. *The Birds of Canada*. Rev. ed. National Museum of Natural Sciences, Ottawa, ON.

Gray, A. P. 1958. *Bird Hybrids: A Check-list with Bibliography*. Technical Communication 13. Commonwealth Agricultural Bureau, Farnham Royal, Bucks, UK.

Haverschmidt, F. 1968. *Birds of Surinam*. Oliver and Boyd, Edinburgh, UK.

Howell, S. N. G., and S. Webb. 1995. *A Guide to the Birds of Mexico and Northern Central America*. Oxford University Press, Oxford, UK.

Hughes, B., and A. J. Green. 2005. Feeding ecology. Pp. 27–56, in J. Kear (ed.). *Ducks, Geese and Swans*, vol. 1. Oxford University Press, Oxford, UK.

Johnsgard, P. A. 1960a. Hybridization in the Anatidae and its taxonomic implications. *Condor* 62: 25–33. http://digitalcommons.unl.edu/biosciornithology/71

Johnsgard, P. A. 1961a. Tracheal anatomy of the Anatidae and its taxonomic significance. *Wildfowl Trust Annual Report* 12: 58–69.

Johnsgard, P. A. 1961b. The taxonomy of the Anatidae—A behavioural analysis. *Ibis* 103a: 71–85. http://digitalcommons.unl.edu/johnsgard/29

Johnsgard, P. A. 1961c. The systematic position of the marbled teal. *Bulletin British Ornithologists' Club* 81: 37–43.

Johnsgard, P. A. 1965. *Handbook of Waterfowl Behavior*. Cornell University Press, Ithaca, NY. http://digitalcommons.unl.edu/bioscihandwaterfowl/7

Johnsgard, P. A. 1968. *Waterfowl: Their Biology and Natural History*. University of Nebraska Press, Lincoln. 138 pp.

Johnsgard, P. A. 1972. Observations on sound production in the Anatidae. *Wildfowl* 22: 46–59. http://digitalcommons.unl.edu/johnsgard/13

Johnsgard, P. A. 1975. *Waterfowl of North America*. Indiana University Press, Bloomington. (Rev. ed. [2010] at http://digitalcommons.unl.edu/biosciwaterfowlna/1)

Johnsgard, P. A. 1978. *Ducks, Geese, and Swans of the World*. University of Nebraska Press, Lincoln. (Rev. ed. [2010] at http://digitalcommons.unl.edu/biosciducksgeeseswans/)

Johnsgard, P. A. 1979a. *Anseriformes* section (Anatidae and Anhimidae). Pp. 425–506, in E. Mayr, ed. *Check-list of the Birds of the World*. Harvard University Press, Cambridge, MA. http://digitalcommons.unl.edu/johnsgard/32

Johnsgard, P. A. 1979b. *A Guide to North American Waterfowl*. Indiana University Press, Bloomington.

Johnsgard, P. A. 1987. *Diving Birds of North America*. University of Nebraska Press, Lincoln. 286 pp.

Johnsgard, P. A. 1992. *Ducks in the Wild: Conserving Waterfowl and Their Habitats*. Key-Porter, Toronto, ON.

Johnsgard, P. A. 2010. *Ducks, Geese, and Swans of the World*. Rev. ed., with a supplement: "The World's Waterfowl in the 21st Century." University of Nebraska–Lincoln DigitalCommons and Zea Books. 498 pp. http://digitalcommons.unl.edu/biosciducksgeeseswans/

Johnsgard, P. A. 2016a. *Swans: Their Biology and Natural History*. University of Nebraska–Lincoln DigitalCommons and Zea Books. 114 pp. http://digitalcommons.unl.edu/zeabook/38

Johnsgard, P. A. 2016b. *The North American Geese: Their Biology and Behavior*. University of Nebraska–Lincoln DigitalCommons and Zea Books. 159 pp. http://digitalcommons.unl.edu/zeabook/44

Johnsgard, P. A. 2016c. *The North American Sea Ducks: Their Biology and Behavior*. University of Nebraska–Lincoln DigitalCommons and Zea Books. 256 pp. http://digitalcommons.unl.edu/zeabook/50

Johnsgard, P. A. 2017a. *The North American Perching and Dabbling Ducks*. University of Nebraska–Lincoln DigitalCommons and Zea Books. 227 pp. http://digitalcommons.unl.edu/zeabook/53/

Johnsgard, P. A., and M. Carbonell. 1996. *Ruddy Ducks and Other Stifftails: Their Behavior and Biology*. University of Oklahoma Press, Norman. 284 pp.

Kear, J., ed. 2005. *Ducks, Geese and Swans*. 2 vols. Oxford University Press, Oxford, UK.

Kear, J., and P. A. Johnsgard. 1968. A review of parental carrying of young by waterfowl. *The Living Bird* 7: 89–102.

Kortright, F. H. 1942. *The Ducks, Geese and Swans of North America*. Wildlife Management Institute, Washington, DC.

Leopold, S. 1959. *Wildlife of Mexico: The Game Birds and Mammals*. University of California Press, Berkeley, CA.

Linduska, J. P., ed. 1964. *Waterfowl Tomorrow*. US Department of the Interior, Bureau of Sport Fisheries and Wildlife, Washington, DC.

Livezey, B. C. 1997. A phylogenetic classification of waterfowl (Aves: Anseriformes), including selected fossil species. *Annals of the Carnegie Museum* 66: 457–496.

Lorenz K. 1971. Comparative studies of the motor patterns of the Anatinae. *Studies in Animal and Human Behavior*. Pp. 14–114. Harvard University Press, London, UK.

Madge, S. 2010. *Wildfowl: An Identification Guide to the Ducks, Geese and Swans of the World*. Bloomsbury Publishing, London, UK.

Madge, S., and H. Burn. 1988. *Waterfowl: An Identification Guide to the Ducks, Geese and Swans of the World*. Houghton Mifflin, Boston, MA.

McKinney, F. 1953. Studies on behaviour of the Anatidae. PhD dissertation, University of Bristol, Bristol, UK.

Millais, I. G. 1913. *The Natural History of British Diving Ducks*. Longmans, Green and Co., London, UK.

Mosby, H. S., ed. 1963. *Wildlife Investigational Techniques*. 2nd ed. Wildlife Society, Washington, DC.

Munroe, B. L. J. 1968. *A Distributional Survey of the Birds of Honduras*. Ornithological Monographs No. 7, American Ornithologists' Union. 758 pp.

Nelson, A. D., and A. C. Martin. 1953. Gamebird weights. *Journal of Wildlife Management* 17: 36–42.

Ogilvie, M. A. 1975. *Ducks of Britain and Europe*. T. & A. D. Poyser, Berkhamsted, UK.

Owen, M. 1977. *Wildfowl of Europe*. Macmillan, London, UK.

Owen, M., and S. Young. 1998. *Wildfowl of the World*. New Holland Publishing, Cape Town, South Africa.

Owen, M., G. L. Atkinson-Willes, and D. Salmon. 1986. *Wildfowl in Great Britain*. 2nd ed. Cambridge University Press, Cambridge, UK.

Palmer, R. S., ed. 1976. *Handbook of North American Birds*, Vol. 2: Waterfowl, Part 1. Yale University Press, New Haven, CT.

Phillips, J. C. 1922–1926. *A Natural History of the Ducks*. 4 vols. Houghton Mifflin, Boston, MA.

Raftovich, R. V., S. C. Chandler, and K. A. Wilkins. 2015. *Migratory Bird Hunting Activity and Harvest during the 2013–14 and 2014–15 Hunting Seasons*. US Fish and Wildlife Service, Laurel, MD.

Reeber, S. 2015. *Wildfowl of Europe, Asia and North America*. Bloomsbury Publishing, London, UK. 655 pp.

Sauer, J. R., W. A. Link, J. E. Fallon, K. L. Pardieck, and D J. Ziolkowski, Jr. 2013. *The North American Breeding Bird Survey 1966–2011: Summary Analysis and Species Accounts*. North American Fauna Number 79: 1–3.

Saunders, G. B, and D. C. Saunders. 1981. *Waterfowl and Their Wintering Grounds in Mexico, 1937–64*. Resource Publication 138. US Fish and Wildlife Service, Washington, DC.

Schneider, K. B. 1965. Growth and plumage development of ducklings in interior Alaska. MS thesis, University of Alaska, Fairbanks.

Todd, F. S. 1979. *Waterfowl: Ducks, Geese and Swans of the World*. Harcourt Brace Jovanovich, New York, and Sea World Press, San Diego, CA.

Todd, F. S. 1996. *Natural History of the Waterfowl*. Ibis Publishing, Vista, CA.

Weller, M. W., ed. 1998. *Waterfowl in Winter*. University of Minnesota Press, Minneapolis.

Wetlands International. 2012. *Waterbird Population Estimates*. 5th ed. Wetlands International, Wageningen, Netherlands.

Wetmore, A. 1965. *The Birds of the Republic of Panama. Pt. 1. Tinamidae (Tinamous) to Rhynchopidae (Skimmers)*. Smithsonian Miscellaneous Collections, Vol. 150. 483 pp.

National, Regional, and Local Surveys

Adamus, P. R., K. Larsen, G. Gillson, and C. R. Miller. 2001. *Oregon Breeding Bird Atlas*. Oregon Field Ornithologists, Eugene.

Addy, C. E. 1964. Atlantic Flyway. Pp. 167–184, in J. P. Linduska, ed. *Waterfowl Tomorrow*. US Department of the Interior, Bureau of Sport Fisheries and Wildlife, Washington, DC.

Alcorn, J. R. 1988. *The Birds of Nevada*. Fairview West Publishing, Fallon, NV.

Aldrich, J. W. 1949. *Migration of Some North American Waterfowl. US Fish and Wildlife Service Special Scientific Report—Wildlife 1*. US Fish and Wildlife Service, Washington, DC.

Andrews, R., and R. Righter. 1992. *Colorado Birds*. Denver Museum of Natural History, Denver, CO.

Bailey, A. M., and R. J. Niedrach. 1965. *Birds of Colorado*. Denver Museum of Natural History, Denver, CO.

Bailey, F. M. 1928. *Birds of New Mexico*. New Mexico Department of Fish and Game, Santa Fe.

Bannerman, D. A. 1958. *Birds of the British Isles*. Vol. 7. Oliver & Boyd, Edinburgh and London, UK.

Bauer, K. M., and U. N. Glutz von Blotzheim. 1968–1969. *Handbuch der Vogel Mitteleuropas*. Vols. 2 and 3. Akademische Verlagsgesellschaft, Frankfurt am Main, Germany.

Baumgartner, F. M., and A. M. Baumgartner. 1992. *Oklahoma Bird Life*. University of Oklahoma Press, Norman.

Benson, K. L. P., and K. A. Arnold. 2001. *The Texas Breeding Bird Atlas*. Texas A&M University System, College Station and Corpus Christi. http://txtbba.tamu.edu

Berrett, D. C. 1962. The birds of the Mexican state of Tabasco. PhD dissertation. Michigan State University, East Lansing.

Bond, J. 1961. *Sixth Supplement to the Check-list of Birds of the West Indies (1956)*. 12 pp. Academy of Natural Sciences, Philadelphia, PA.

Bond, J. 1971. *Birds of the West Indies*. 2nd ed. Houghton Mifflin, Boston, MA.

Brown, D. E. 1985. *Arizona Wetlands and Waterfowl*. University of Arizona Press, Tucson.

Brown, M. B., and P. A. Johnsgard. 2013. *Birds of the Central Platte River Valley and Adjacent Counties*. University of Nebraska–Lincoln DigitalCommons and Zea Books. 182 pp. http://digitalcommons.unl.edu/zeabook/15

Burleigh, T. D. 1944. *The Bird Life of the Gulf Coast Region of Mississippi*. Louisiana State University, Museum of Zoology Occasional Papers, No. 20, pp. 329–490.

Burleigh, T. D. 1972. *Birds of Idaho*. Caxton Printers Ltd., Caldwell, ID.

Byrd, G. V., D. L. Johnson, and D. D. Gibson. 1974. The birds of Adak Island, Alaska. *Condor* 76: 288–300.

Cadman, M. D., P. J. F. Engels, and F. M. Helleiner, eds. 2016. *Atlas of the Breeding Birds of Ontario*. 2nd ed. Federation of Ontario Naturalists and Long Point Bird Observatory, Bird Studies Canada, Environment Canada, Ontario Field Ornithologists, Ontario Ministry of Natural Resources, and Ontario Nature, Toronto.

Campbell, R. W., N. K. Dawe, I. McTaggart-Cowan, J. M. Cooper, G. W. Kaiser, and M. C. E. McNall. 1990. *The Birds of British Columbia, Vol. 1: Nonpasserines, Introduction and Loons through Waterfowl.* University of British Columbia Press, Vancouver.

Canterbury, J., P. A. Johnsgard, and H. Downing. 2013. *Birds and Birding in Wyoming's Bighorn Mountains Region.* University of Nebraska–Lincoln and Zea Books. 260 pp. http://digitalcommons.unl.edu/zeabook/18

Castrale, J. S., E. M. Hopkins, and C. E. Keller. 1998. *Atlas of Breeding Birds of Indiana.* Indiana Department of Natural Resources, Indianapolis.

Chamberlain, E. B. 1960. *Florida Waterfowl Populations, Habitats and Managements.* Florida Game and Fresh Water Fish Commission, Technical Bulletin No. 7. 62 pp.

Cogswell, H. L. 1977. *Water Birds of California.* University of California Press, Berkeley.

Corman, T. E., and C. Wise-Gervais. 2005. *Arizona Breeding Bird Atlas.* University of New Mexico Press, Albuquerque.

Davidson, P. J. A., R. J. Cannings, A. R. Couturier, D. Lepage, and C. M. D. Corrado, eds. 2008 et seq. *The Atlas of the Breeding Birds of British Columbia.* Bird Studies Canada, Delta, BC. (Species accounts are available at http://www.birdatlas.bc.ca/accounts/toc.jsp?show=species)

Erskine, A. J. 2016. *Atlas of Breeding Birds of the Maritime Provinces.* 2nd ed. Bird Studies Canada, Sackville, NS.

Gabrielson, I. N., and F. C. Lincoln. 1959. *Birds of Alaska.* Stackpole, Harrisburg, PA, and Wildlife Management Institute, Washington, DC.

Gauthier, J., and Y. Aubry, eds. 1996. *The Breeding Birds of Québec: Atlas of the Breeding Birds of Southern Québec.* Province of Québec Society for the Protection of Birds, Canadian Wildlife Service, Québec Region, Montréal.

Gibson, D. D., and G. V. Byrd. 2007. *Birds of the Aleutian Islands, Alaska.* Nuttall Ornithological Club, Cambridge, MA, and American Ornithologists' Union, Washington, DC.

Gilligan, J., M. Smith, D. Rogers, and A. Contreras, eds. 1994. *Birds of Oregon: Status and Distribution.* Cinclus Publications, McMinnville, OR.

Godfrey, W. E. 1986. *The Birds of Canada.* Rev. ed. National Museum of Natural Sciences, Ottawa, ON.

Grinnell, J., and A. H. Miller. 1944. *The Distribution of the Birds of California.* Pacific Coast Avifauna 27. Cooper Ornithological Club, Berkeley, CA.

Griscom, L., and D. E. Snyder. 1955. *The Birds of Massachusetts: An Annotated and Revised Check List.* Peabody Museum, Salem, MA.

Gromme, O. J. 1963. *Birds of Wisconsin.* University of Wisconsin Press, Madison.

Haverschmidt, F. 1968. *Birds of Surinam.* Oliver and Boyd, Edinburgh, UK.

Jackson, L. S., C. A. Thompson, and J. J. Dinsmore. 1996. *The Iowa Breeding Bird Atlas.* University of Iowa Press, Iowa City.

Janssen, R. B. 1987. *Birds in Minnesota.* University of Minnesota Press, Minneapolis.

Johnsgard, P. A. 1968. *Waterfowl: Their Biology and Natural History.* University of Nebraska Press, Lincoln. 138 pp.

Johnsgard, P. A. 1986. *Birds of the Rocky Mountains, with Particular Reference to National Parks in the Northern Rocky Mountain Region.* Colorado Associated University Press, Boulder, CO. 504 pp.

Johnsgard, P. A. 1997. *The Avian Brood Parasites: Deception at the Nest.* Oxford University Press, New York. 409 pp.

Johnsgard, P. A. 2009. *Birds of the Great Plains: Breeding Species and Their Distribution.* University of Nebraska Press, Lincoln. Rev. ed. with a 2009 supplement: http://digitalcommons.unl.edu/bioscibirdsgreatplains

Johnsgard, P. A. 2011. *Rocky Mountain Birds: Birds and Birding in the Central and Northern Rockies.* University of Nebraska–Lincoln DigitalCommons and Zea Books. 274 pp. http://digitalcommons.unl.edu/zeabook/7

Johnsgard, P. A. 2012a. *Wings over the Great Plains: Bird Migrations in the Central Flyway.* University of Nebraska–Lincoln and Zea Books. 249 pp. http://digitalcommons.unl.edu/zeabook/13

Johnsgard, P. A. 2012b. *Wetland Birds of the Central Plains: South Dakota, Nebraska and Kansas.* University of Nebraska–Lincoln DigitalCommons and Zea Books. 275 pp. http://digitalcommons.unl.edu/zeabook/8

Johnsgard, P. A. 2013. *The Birds of Nebraska.* Rev. ed. University of Nebraska–Lincoln and Zea Books. 150 pp. http://digitalcommons.unl.edu/zeabook/17

Johnsgard, P. A. 2015. *Global Warming and Population Responses among Great Plains Birds.* University of Nebraska–Lincoln DigitalCommons and Zea Books. 384 pp. http://digitalcommons.unl.edu/zeabook/26

Johnston, R. F. 1964. The breeding birds of Kansas. *University of Kansas Publications of the Museum of Natural History* 12: 575–655.

Keith, L. B. 1961. *A Study of Waterfowl Ecology on Small Impoundments in Southeastern Alberta*. Wildlife Monographs, No. 6. American Ornithologists' Union, Washington, DC.

Kessel, B. 1989. *Birds of the Seward Peninsula, Alaska: Their Biogeography, Seasonality, and Natural History*. University of Alaska Press, Fairbanks.

Kessel, B., and D. G. Gibson. 1978. *Status and Distribution of Alaska Birds*. Studies in Avian Biology 1. Cooper Ornithological Society, Los Angeles, CA.

Kingery, H. E., ed. 1997. *Colorado Breeding Bird Atlas*. Colorado Bird Partnership, Denver.

Koskimies, J., and L. Lahti. 1964. Cold-hardiness of the newly hatched young in relation to ecology and distribution in ten species of European ducks. *Auk* 81: 281–307.

Lee, F. B., R. L. Jessen, N. J. Ordal, R. I. Benson, J. P. Lindmeier, and L. L. Johnson. 1964. *Waterfowl in Minnesota*. Technical Bulletin 7. Minnesota Department of Conservation, St. Paul.

Larrison, E. J., and K. G. Sonnenburg. 1968. *Washington Birds: Their Location and Identification*. Seattle Audubon Society, Seattle, WA.

Lockwood, M. W., and B. Freeman. 2014. *Handbook of Texas Birds*. 2nd ed. Texas A&M University Press, College Station.

Marks, J., P. Hendricks, and D. Casey. 2016. *Birds of Montana*. Buteo Books, Arlington, VA.

Martin, A. C., H. S. Zim, and A. L. Nelson. 1951. *American Wildlife and Plants*. McGraw-Hill, New York.

McGowan, K. J., and K. Corwin. 2008. *The Second Atlas of Breeding Birds in New York State*. Cornell University Press, Ithaca, NY.

McPeek, G. A., ed. 1994. *The Birds of Michigan*. Indiana University Press, Bloomington.

McWilliams, G. M., and D. W. Brauning. 2000. *The Birds of Pennsylvania*. Cornell University Press, Ithaca, NY.

Moyle, J. B., F. B. Lee, R. L. Jessen, N. J. Ordal, R. I. Benson, J. P. Lindmeier, and L. L. Johnson. 1964. *Waterfowl in Minnesota*. Technical Bulletin 7. Division of Game and Fish, Minnesota Department of Conservation, St. Paul.

Munroe, B. L. J. 1968. *A Distributional Survey of the Birds of Honduras*. Ornithological Monographs No. 7, American Ornithologists' Union. 758 pp.

Murie, O. J. 1959. *Fauna of the Aleutian Islands and Alaska Peninsula*. US Department of Interior, Fish and Wildlife Service, North American Fauna, No. 61, pp. 1–406.

Musgrove, J. W., and M. R. Musgrove. 1947. *Waterfowl in Iowa*. 2nd ed. State Conservation Commission, Des Moines, IA.

Oakleaf, B., B. Luce, S. Ritter, and A. Cerovski, eds. 1992. *Wyoming Bird and Mammal Atlas*. Wyoming Game and Fish Department, Lander.

Oberholser, H. C. 1974. *The Bird Life of Texas*. Vol. 1. University of Texas Press, Austin.

Palmer, R. S. 1949. *Maine Birds*. Bulletin of the Museum of Comparative Zoology 102. Museum of Comparative Zoology, Cambridge, MA.

Peterjohn, B. G. 1989. *The Birds of Ohio*. Indiana University Press, Bloomington.

Peterjohn, B. G., and D. L. Rice. 1991. *The Ohio Breeding Bird Atlas*. Ohio Department of Natural Resources, Columbus.

Peters, H. S., and T. D. Burleigh. 1951. *The Birds of Newfoundland*. Newfoundland Department of Natural Resources, St. Johns.

Peterson, R. A. 1995. *The South Dakota Breeding Bird Atlas*. South Dakota Ornithologists' Union, Aberdeen.

Phillips, A. R., J. Marshall, and G. Monson. 1964. *Birds of Arizona*. University of Arizona Press, Tucson.

Potter, E. F., J. F. Parnell, and R. D. Teulings. 1980. *Birds of the Carolinas*. University of North Carolina Press, Chapel Hill.

Pranty, B., K. A. Radamaker, and G. Kennedy. 2006. *Birds of Florida*. Lone Pine Publishing International, Auburn, WA.

Reinking, D. A., ed. 2004. *Oklahoma Breeding Bird Atlas*. University of Oklahoma Press, Norman.

Roberts, T. R. 1932. *The Birds of Minnesota*. 2 vols. University of Minnesota Press, Minneapolis.

Robbins, M. B., and D. A. Easterla. 1992. *Birds of Missouri: Their Distribution and Abundance*. University of Missouri Press, Columbia.

Robbins, S. D., Jr., 1991. *Wisconsin Birdlife*. University of Wisconsin Press, Madison.

Rodner, C., M. Lentino, and R. Restall. 2000. *Checklist of the Birds of Northern South America*. Yale University Press, New Haven, CT.

Salomonsen, F. 1950. *The Birds of Greenland*. Part 1. Ejnar Munksgaard, Copenhagen, Denmark.

Salt, W. R., and J. R. Salt. 1976. *The Birds of Alberta*. Hurtig Publishers, Edmonton, AB.

Schiøler, E. 1925–1926. *Danmarks Fugle*. 2 vols. Gyldendelske, Denmark.

Semenchuk, G. P. 1992. *The Atlas of Breeding Birds of Alberta*. Federation of Alberta Naturalists, Edmonton, AB.

Singleton, J. R. 1953. *Texas Coastal Waterfowl Survey*. Texas Game and Fish Commission, F.A. Report, Series 11. 128 pp.

Small, A. 1994. *California Birds: Their Status and Distribution*. Ibis Publishing, Temecula, CA.

Smith, A. G. 1971. *Ecological Factors Affecting Waterfowl Production in the Alberta Parklands*. Resource Publication 98. US Department of the Interior, Fish and Wildlife Service, Bureau of Sport Fisheries and Wildlife, Washington, DC.

Smith, A. R. 1996. *Atlas of Saskatchewan Birds*. Saskatchewan Natural History Society, Regina.

Snyder, L. L. 1957. *Arctic Birds of Canada*. University of Toronto Press, Toronto, ON.

Sprunt, A., Jr., and E. B. Chamberlain. 1949. *South Carolina Bird Life*. University of South Carolina Press, Columbia.

Stevenson, H. M., and B. H. Anderson. 1994. *The Birdlife of Florida*. University Press of Florida, Gainesville.

Stewart, R. E. 1962. *Waterfowl Populations in the Upper Chesapeake Region*. US Department of Interior, Fish and Wildlife Service, Bureau of Sport Fisheries and Wildlife, Special Scientific Report—Wildlife No. 65. 208 pp.

Stewart, R. E. 1975. *Breeding Birds of North Dakota*. Tri-College Center for Regional Studies, Fargo, ND.

Stiles, F. G., and A. F. Skutch. 1989. *A Guide to the Birds of Costa Rica*. Cornell University Press, Ithaca, NY.

Sugden, L. G. 1973. *Feeding Ecology of Pintail, Gadwall, American Wigeon and Lesser Scaup Ducklings in Southern Alberta*. Canadian Wildlife Service Report Series, No. 24. 32 pp.

Sutton, G. M. 1967. *Oklahoma Birds*. University of Oklahoma Press, Norman.

Texas Game, Fish and Oyster Commission. 1945. *Principal Game Birds and Mammals of Texas*. Texas Game, Fish and Oyster Commission, Austin.

Todd, W. E. C. 1963. *Birds of the Labrador Peninsula and Adjacent Areas*. University of Toronto Press, Toronto, ON.

Townsend, G. H. 1966. A study of waterfowl nesting on the Saskatchewan River delta. *Canadian Field-Naturalist* 80: 74–88.

Tufts, R. W. 1986. *Birds of Nova Scotia*. 3rd ed. Nimbus Publishing and Nova Scotia Museum, Halifax, NS.

Turcotte, W. H., and D. L. Watts. 1999. *Birds of Mississippi*. University Press of Mississippi, Jackson.

US Fish and Wildlife Service. 2016. *Waterfowl Population Status, 2016*. U. S. Department of the Interior, Washington, DC. https://www.fws.gov/migratorybirds/pdf/surveys-and-data/Population-status/Waterfowl/WaterfowlPopulationStatusReport16.pdf

US Geological Survey. 2013. *North American Breeding Bird Survey, Results and Analysis 1966–2013*. Patuxent Wildlife Research Center, Laurel, MD. http://www.mbr-pwrc.usgs.gov/bbs/bbs2013.html

Veit, R. R., and W. R. Petersen. 1993. *Birds of Massachusetts*. Massachusetts Audubon Society, Lincoln.

Wahl, T. R., B. Tweit, and S. G. Mlodinow. 2005. *Birds of Washington: Status and Distribution*. Oregon State University Press, Corvallis.

Wetmore, A. 1965. *The Birds of the Republic of Panama. Pt. 1. Tinamidae (Tinamous) to Rhynchopidae (Skimmers)*. Smithsonian Miscellaneous Collections, Vol. 150. 483 pp.

Wick, W. Q., and R. G. Jeffrey. 1966. Population estimates and hunter harvest of diving ducks in northeastern Puget Sound, Washington. *Murrelet* 47: 23–32.

Wiedenfeld, D. A., and M. M. Swan. 2000. *Louisiana Breeding Bird Atlas*. Louisiana Sea Grant College Program, Baton Rouge.

Williams, C. S., and W. H. Marshall. 1938. Duck nesting studies, Bear River Migratory Bird Refuge, Utah, 1937. *Journal of Wildlife Management* 2: 29–48.

Yocom, C. F. 1951. *Waterfowl and Their Food Plants in Washington*. University of Washington Press, Seattle.

Zimmerman, D. A., and J. Van Tyne. 1959. A distributional check-list of the birds of Michigan. *University of Michigan Museum of Zoology Occasional Papers*, No. 608: 1–63.

Zimpfer, N., W. E. Rhodes, D. D. Silverman, G. S. Zimmerman, and K. D. Richkus. 2014. *Trends in Duck Breeding Populations, 1955–2014*. US Fish and Wildlife Service, Laurel, MD.

Whistling-Ducks: Multiple Taxa

Bolen, E. G., and M. K. Rylander. 1983. *Whistling-Ducks: Zoogeography, Ecology, Anatomy.* Special Publication 20, The Museum, Texas Tech University Press, Lubbock.

Bruzual, J., and I. Bruzual. 1983. Feeding habits of whistling ducks in the Calabozo ricefields, Venezuela, during the nonreproductive period. *Wildfowl* 34: 20–26.

Callighan, D. 2005. The whistling-ducks. Pp. 187–189, in J. Kear, ed. *Ducks, Geese and Swans.* Vol. 1. Oxford University Press, Oxford, UK.

Clark, A. 1976. Observations on the breeding of whistling ducks in southern Africa. *Ostrich* 47: 59–64.

Clark, A. 1978. Some aspects of the behaviour of whistling ducks in South Africa. *Ostrich* 49: 31–39.

Dallmeier, F. 1991. Whistling-ducks as a manageable and sustainable resource in Venezuela: Balancing economic costs and benefits. Pp. 266–287 in J. G. Robinson and K. H. Redford, eds. *Neotropical Wildlife Use and Conservation.* University of Chicago Press, Chicago.

Feekes, F., M. Morales M., and O. Sorianos S. 1992. Nest boxes made from native materials for whistling-ducks. *Wildlife Society Bulletin* 20: 113–115.

Johnstone, S. T. 1957. On breeding of whistling ducks. *Avicultural Magazine* 63: 23–25.

Rylander, M. K., and E. G. Bolen. 1970. Ecological and anatomical adaptations of North American tree ducks. *Auk* 87: 72–90.

Rylander, M. K., and E. G. Bolen. 1974a. Analysis and comparison of gaits in whistling-ducks (*Dendrocygna*). *Wilson Bulletin* 86: 237–245.

Rylander, M. K., and E. G. Bolen. 1974b. Feeding adaptations in whistling-ducks (*Dendrocygna*). *Auk* 91: 86–94.

Rylander, M. K., E. G. Bolen, and R. E. McCamant. 1980. Evidence of incubation patches in whistling-ducks. *Southwestern Naturalist* 25: 126–128.

Fulvous Whistling-Duck

Acosta, M., and L. Mugica. 2006a. Aves en el ecosistema arrocero. Pp. 108–135 in L. Mugica, D. Denis, M. Acosta, A. Jiménez, and A. Rodríguez, eds. *Aves Acquáticas en los Humedales de Cuba.* Editorial Científico-Técnica, Havana, Cuba.

Acosta, M., and L. Mugica. 2006b. *Reporte Final de Aves Acuáticas en Cuba.* Facultad de Biología, Universidad de La Habana, Havana, Cuba.

Acosta Cruz, M., L. Mugica Valdés, and O. Tórres Fundora. 1989. Ecomorfología de *Dendrocygna bicolor* (Vieillot) (Aves: Anatidae) en Cuba. *Ciencias Biológicas* 21/22: 70–78.

Carroll, J. J. 1932. A change in the distribution of the fulvous tree duck in Texas. *Auk* 49: 343–344.

Cottam, C., and W. C. Glazener. 1959. Late nesting of water birds in south Texas. *Transactions of the North American Wildlife Conference* 24: 382–394.

Dickey, D. R., and A. J. Van Rossem. 1923. The fulvous tree-ducks of Buena Vista Lake. *Condor* 25: 39–50.

Flickinger, E. L., and K. A. King. 1972. Some effects of aldrin-treated rice on Gulf Coast wildlife. *Journal of Wildlife Management* 36: 706–727.

Flickinger, E. L., D. S. Lobpries, and H. A. Bateman. 1977. Fulvous whistling-duck populations in Texas and Louisiana. *Wilson Bulletin* 89: 329–331.

Flickinger, E. L., K. A. King, and O. Heyland. 1973. Pen-reared fulvous tree ducks used in movement studies of wild populations. *Journal of Wildlife Management* 37: 171–175.

Gerstenberg, G., and C. D. Rey. 2004. Record of nesting by fulvous whistling-ducks, southern San Joaquin Valley, California. *California Fish and Game* 90: 155–156.

Gómez Ventura, J. A., and Z. R. de Mendoza. 1982. Aspectos generals sobre la reproducción del piche real, *Dendrocygna bicolor,* en la Laguna del Jocotal, El Salvador. *Actas del Congreso Latinoamericano de Zoología* 8: 807–820.

Hamilton, R. A. 2008. Fulvous whistling-duck (*Dendrocygna bicolor*). *Studies of Western Birds* 1: 68–73.

Hartz, S. T. 1962. The distribution of the fulvous tree duck. *Cassinia* 46: 10–12.

Hohman, W. L., and D. M. Richard. 1994. Timing of remigial molt in fulvous whistling-ducks nesting in Louisiana. *Southwestern Naturalist* 39: 190–192.

Hohman, W. L., J. L. Moore, T. M. Stark, G. A. Weisbrich, and R. A. Coon. 1994. Breeding waterbird use of Louisiana rice fields in relation to planting practices. *Proceedings of the Southeastern Association of Fish and Wildlife Agencies* 48: 31–37.

Hohman, W. L., T. M. Stark, and J. L. Moore. 1996. Food availability and feeding preferences of breeding fulvous whistling-ducks in Louisiana rice fields. *Wilson Bulletin* 108: 137–150.

Jarrett, N. 2005. Fulvous whistling-duck. Pp. 199–202, in J. Kear, ed. *Ducks, Geese and Swans*. Vol. 1. Oxford University Press, Oxford, UK.

Jones, H. L. 1966. The fulvous tree duck in the East: Its past and present status. *Chat* 30: 4–7.

Lynch, J. J. 1943. Fulvous tree duck in Louisiana. *Auk* 60: 100–102.

McCartney, R. B. 1963. The fulvous tree duck in Louisiana. MS thesis, Louisiana State University, Baton Rouge.

Meanley, B., and A. G. Meanley. 1958. Post-copulatory display in fulvous and black-bellied tree ducks. *Auk* 75: 96.

Meanley, B., and A. G. Meanley. 1959. Observations on the fulvous tree duck in Louisiana. *Wilson Bulletin* 71: 33–45.

Munro, W. T. 1967. Occurrence of the fulvous tree duck in Canada. *Canadian Field-Naturalist* 81: 151–152.

Peris, S. J., B. Sánchez, and D. Rodríguez. 1998. Range expansion of the fulvous whistling-duck (*Dendrocygna bicolor*) in Cuba in relation to rice cultivation. *Caribbean Journal of Science* 34: 164–166.

Shields, A. M. 1899. Nesting of the fulvous tree duck. *Cooper Ornithological Club Bulletin* 1: 9–11.

Turnbull, R. E., F. A. Johnson, and D. H. Brakhage. 1989. Status, distribution, and foods of fulvous whistling-ducks in south Florida. *Journal of Wildlife Management* 53: 1046–1051.

Turnbull, R. E., F. A. Johnson, M. A. Hernandez, W. B. Wheeler, and J. P. Toth. 1989. Pesticide residues in fulvous whistling-ducks from south Florida. *Journal of Wildlife Management* 53: 1052–1057.

Tweit, R. C. 2008. Fulvous whistling-duck. In *The Texas Breeding Bird Atlas*. http://txtbba.tamu.edu/species-accounts/fulvous-whistling-duck

Volodin, I., M. Kaiser, M. Martin, V. Matrosova, E. Volodina, A. Klenova, O. Filatova, and M. Kholodova. 2009. The technique of noninvasive distant sexing for four monomorphic *Dendrocygna* whistling duck species by their loud whistles. *Bioacoustics* 18: 277–290.

Wyss, A. J. 1996. Nesting ecology of fulvous whistling-ducks in the Everglades agricultural area of southern Florida. MS thesis, Auburn University, Auburn, AL.

Zwank, P. J., P. M. McKenzie, and E. B. Moser. 1988. Fulvous whistling-duck abundance and habitat use in southwestern Louisiana. *Wilson Bulletin* 100: 488–494.

West Indian Whistling-Duck

Bond, J. 1971. *Birds of the West Indies*. 2nd ed. Houghton Mifflin, Boston, MA.

Bradley, P. E. 2000. *The Birds of the Cayman Islands*. Checklist No. 19. British Ornithologists' Union, Tring, UK.

Staus, N. 1998a. Habitat and home range of West Indian whistling-duck. *Journal of Wildlife Management* 62: 17–178.

Staus, N. 1998b. Behavior and natural history of the West Indian whistling-duck on Long Island, Bahamas. *Wildfowl* 49: 194–206.

Staus, N. 2005. West Indian whistling-duck. Pp. 197–199, in J. Kear, ed. *Ducks, Geese and Swans*. Vol. 1. Oxford University Press, Oxford, UK.

Todd, F. S. 1979. *Waterfowl: Ducks, Geese and Swans of the World*. Harcourt Brace Jovanovich, New York, and Sea World Press, San Diego, CA.

Todd, F. S. 1996. *Natural History of the Waterfowl*. Ibis Publishing, Vista, CA. (West Indian whistling-duck, p. 81.)

Black-bellied Whistling-Duck

Ballard, B. M. 2001. Parasitism of a laughing gull nest by black-bellied whistling-ducks. *Wilson Bulletin* 113: 339–340.

Banks, R. C. 1978. Nomenclature of the black-bellied whistling-duck. *Auk* 95: 348–352.

Bergman, D. L. 1994. Post-hatch brood amalgamation by black-bellied whistling-ducks. *Wilson Bulletin* 106: 563–564.

Bergstrom, B. J. 1999. First reported breeding of black-bellied whistling-duck in northern Florida. *Florida Field Naturalist* 27: 177–179.

Bolen, E. G. 1964. Weights and linear measurements of black-bellied tree ducks. *Texas Journal of Science* 16: 257–260.

Bolen, E. G. 1967a. The ecology of the black-bellied tree duck in southern Texas. PhD dissertation, Utah State University, Logan.

Bolen, E. G. 1967b. Nesting boxes for black-bellied tree ducks. *Journal of Wildlife Management* 31: 794–797.

Bolen, E. G. 1970. Sex ratios in the black-bellied tree duck. *Journal of Wildlife Management* 34: 68–73.

Bolen, E. G. 1971. Pair-bond tenure in the black-bellied tree duck. *Journal of Wildlife Management* 35: 385–388.

Bolen, E. G. 2005. Black-bellied whistling-duck. Pp. 192–195, in J. Kear, ed. *Ducks, Geese and Swans.* Vol. 1. Oxford University Press, Oxford, UK.

Bolen, E. G., and B. J. Forsyth. 1967. Foods of the black-bellied tree duck in south Texas. *Wilson Bulletin* 79: 43–49.

Bolen, E. G., and B. W. Cain. 1968. Mixed wood duck–tree duck clutch in Texas. *Condor* 70: 389–390.

Bolen, E. G., and E. N. Smith. 1979. Notes on the incubation behavior of black-bellied whistling-ducks. *Prairie Naturalist* 11: 119–123.

Bolen, E. G., and J. J. Beecham. 1970. Notes on the foods of juvenile black-bellied tree ducks. *Wilson Bulletin* 82: 325–326.

Bolen, E. G., and R. E. McCamant. 1977. Mortality rates for black-bellied whistling-ducks. *Bird-Banding* 48: 350–353.

Bolen, E. G., B. McDaniel, and C. Cottam. 1964. Natural history of the black-bellied tree duck (*Dendrocygna autumnalis*) in southern Texas. *Southwestern Naturalist* 9: 78–88.

Bourne, G. R. 1979. Weights and linear measurements of black-bellied whistling-ducks in Guyana. *Welder Wildlife Foundation Symposium* 1: 186–188.

Bourne, G. R. 1981. Food habits of black-bellied whistling-ducks occupying rice culture habitats. *Wilson Bulletin* 93: 551–554.

Bourne, G. R., and D. R. Osborne. 1978. Black-bellied whistling-duck utilization of a rice culture habitat. *Interciencia* 3: 152–159.

Burgess, H. H. 2007. Black-bellied whistling-duck. In *The Texas Breeding Bird Atlas.* http://txtbba.tamu.edu/species-accounts/black-bellied-whistling-duck

Cain, B. W. 1968. Growth and plumage development of the black-bellied tree duck, *Dendrocygna autumnalis* (Linnaeus). MS thesis, Texas A&I University, Kingsville.

Cain, B. W. 1970. Growth and plumage development of the black-bellied tree duck, *Dendrocygna autumnalis. Texas A&I University Studies* 3: 25–48.

Cain, B. W. 1973. Effect of temperature on energy requirements and northward distribution of the black-bellied tree duck. *Wilson Bulletin* 85: 308–317.

Cain, B. W. 1976. Energetics of growth for black-bellied tree ducks. *Condor* 78: 124–128.

Chronister, C. D. 1985. Egg-laying and incubation behavior of black-bellied whistling-ducks. MS thesis, University of Minnesota–St. Paul, St. Paul, MN.

Delnicki, D. 1983. Mate changes by black-bellied whistling-ducks. *Auk* 100: 728–729.

Delnicki, D., and E. G. Bolen. 1975. Natural nest site availability for black-bellied whistling-ducks in south Texas. *Southwestern Naturalist* 20: 371–378.

Delnicki, D., and E. G. Bolen. 1976. Renesting by the black-bellied whistling-duck. *Auk* 93: 535–542.

Delnicki, D. E. 1973. Renesting, incubation behavior, and compound clutches of the black-bellied tree duck in southern Texas. MS thesis, Texas Tech University, Lubbock.

Dozier, J. 2012. Black-bellied whistling-duck nest in the Santee Delta-Winyah area of South Carolina. *Chat* 76: 16–18.

Feekes, F. 1991. The black-bellied whistling-duck in Mexico: From traditional use to sustainable management? *Biological Conservation* 56: 123–131.

Harrigal, D., and J. E. Cely. 2004. Black-bellied whistling-ducks nest in South Carolina. *Chat* 68: 106–108.

Harrigal, D., P. Laurie, and C. Floyd. 1995. Black-bellied whistling-ducks observed in Colleton County, South Carolina. *Chat* 59: 99–100.

Heins, M. S. 1984. Ecology and behavior of black-bellied whistling-duck broods in south Texas. MS thesis, Texas Tech University, Lubbock.

Heins-Loy, M. 1986a. Brood mortality of black-bellied whistling-ducks in south Texas. *Journal of Field Ornithology* 57: 233–235.

Heins-Loy, M. 1986b. Fall age ratios of the black-bellied whistling-duck. *Southwestern Naturalist* 31: 107–109.

Hersloff, L., P. N. Lehner, E. G. Bolen, and M. K. Rylander. 1974. Visual sensitivity in the black-bellied tree duck (*Dendrocygna autumnalis*), a crepuscular species. *Journal of Comparative and Physiological Psychology* 86: 486–492.

James, J. D. 2000. Effects of habitat and spatial characteristics on the incidence of conspecific brood parasitism and nest site selection in breeding black-bellied whistling-ducks. MS thesis, Texas A&M University, Kingsville.

James, J. D., J. E. Thompson, and B. M. Ballard. 2012. Evidence of double brooding by black-bellied whistling-ducks. *Wilson Journal of Ornithology* 124: 183–185.

Johnson, A. R., and J. C. Barlow. 1971. Notes on the nesting of the black-bellied tree duck near Phoenix, Arizona. *Southwestern Naturalist* 15: 394–395.

Kamp, M. B., and J. Loyd. 2001. First breeding record of the black-bellied whistling-duck for Oklahoma. *Bulletin of the Oklahoma Ornithological Society* 34: 13–17.

Kramer, G. W., and N. H. Euliss. 1986. Winter foods of black-bellied whistling-ducks in northwestern Mexico. *Journal of Wildlife Management* 50: 413–416.

Markum, D. E., and G. A. Baldassarre. 1989. Ground nesting by black-bellied whistling-ducks on islands in Mexico. *Journal of Wildlife Management* 53: 707–713.

McCamant, R. E., and E. G. Bolen. 1979. A 12-year study of nest box utilization by black-bellied whistling-ducks. *Journal of Wildlife Management* 43: 936–943.

Meanley, B., and A. G. Meanley. 1958a. Nesting habitat of the black-bellied tree duck in Texas. *Wilson Bulletin* 70: 94–95.

Meanley, B., and A. G. Meanley. 1958b. Post-copulatory display in fulvous and black-bellied tree ducks. *Auk* 75: 96.

Schneider, J. P., T. C. Tacha, and D. Lobpries. 1993. Breeding distribution of black-bellied whistling-ducks in Texas. *Southwestern Naturalist* 38: 383–385.

Pochards: Multiple Taxa

Austin, J. E., A. D. Afton, M. G. Anderson, R. G. Clark, C. M. Custer, J. S. Lawrence, J. B. Pollard, and J. K. Ringelman. 2000. Declining scaup populations: Issues, hypotheses, and research needs. *Wildlife Society Bulletin* 28: 254–263.

Austin, J. E., coordinator. 2010. *Conservation Action Plan for Greater Scaup and Lesser Scaup, Version 1.0.* US Geological Survey, Northern Prairie Wildlife Research Center, Jamestown, ND, and Division of Migratory Bird Management, US Fish and Wildlife Service, Patuxent, MD, and Region 3, Minneapolis, MN.

Badzinski, S. S., and S. A. Petrie. 2006. Diets of lesser and greater scaup during autumn and spring on the lower Great Lakes. *Wildlife Society Bulletin* 34: 664–674.

Bailey, R. O. 1983. *Distribution of Postbreeding Diving Ducks (Aythyini and Mergini) on Southern Boreal Lakes in Manitoba.* Progress Notes 136. Canadian Wildlife Service, Ottawa, ON.

Bartonek, J. C., and J. J. Hickey. 1969. Food habits of canvasbacks, redheads, and lesser scaup in Manitoba. *Condor* 71: 280–290.

Bergman, R. D. 1973. Use of southern boreal lakes by postbreeding canvasbacks and redheads. *Journal of Wildlife Management* 37: 160–170.

Bouffard, S. H. 1983. Redhead egg parasitism of canvasback nests. *Journal of Wildlife Management* 47: 213–216.

Callaghan, D. 2005. The pochards and scaup (diving ducks). Pp. 621–624, in J. Kear, ed. *Ducks, Geese and Swans.* Vol. 2. Oxford University Press, Oxford.

Campbell, J. M. 1969. The canvasback, common goldeneye and bufflehead in arctic Alaska. *Condor* 71: 80.

Cottam, C. 1939. *Food Habits of North American Diving Ducks*. Technical Bulletin 643. US Department of Agriculture, Washington, DC.

Cronan, J. M. 1957. Food and feeding habits of the scaups in Connecticut waters. *Auk* 74: 459–468.

Custer, C. M., and T. W. Custer. 1996. Food habits of diving ducks in the Great Lakes after the zebra mussel invasion. *Journal of Field Ornithology* 67: 86–99.

Eccleston, K. A. 1999. Food sources as a factor in the decline of greater scaup and lesser scaup ducks. MS thesis, University of Connecticut, Storrs.

Fournier, M. A., and J. E. Hines. 1996. Nest sharing by a lesser scaup and a greater scaup. *Wilson Bulletin* 108: 380–381.

Fournier, M. A., and J. E. Hines. 2001. Breeding ecology of sympatric greater and lesser scaup (*Aythya marila* and *Aythya affinis*) in the subarctic Northwest Territories. *Arctic* 54: 444–456.

Gammonley, J. H., and M. E. Heitmeyer. 1990. Behavior, body condition, and foods of buffleheads and lesser scaups during spring migration through the Klamath Basin, California. *Wilson Bulletin* 102: 672–683.

Henny, C. J. 1970. *Winter Bandings of Mallards, Black Ducks, Wood Ducks, Pintails, Canvasbacks, Redheads, and Scaup in 1967, 1968, and 1969*. Administrative Report 197. US Bureau of Sport Fisheries and Wildlife, Laurel, MD.

Hildén, O. 1964. Ecology of duck populations in the island group of Valassaaret, Gulf of Bothnia. *Annales Zoologici Fennici* 1: 1–279.

Jarvis, R. L., and J. H. Noyes. 1986. Foods of canvasbacks and redheads in Nevada: Paired males and ducklings. *Journal of Wildlife Management* 50: 199–203.

Kessel, B., and D. G. Gibson. 1978. *Status and Distribution of Alaska Birds*. Studies in Avian Biology 1. Cooper Ornithological Society, Los Angeles, CA.

Kinney, S. D. 2004. Estimating the population of greater and lesser scaup during winter in off-shore Louisiana. MS thesis, Louisiana State University, Baton Rouge.

Koons, D. N., and J. J. Rotella. 2003. Comparative nesting success of sympatric lesser scaup and ring-necked ducks. *Journal of Field Ornithology* 74: 222–229.

Lagerquist, B. A., and C. D. Ankney. 1989. Interspecific differences in bill and tongue morphology among diving ducks (*Aythya* spp., *Oxyura jamaicensis*). *Canadian Journal of Zoology* 67: 2694–2699.

Lightbody, J. P., and C. D. Ankney. 1984. Seasonal influence on the strategies of growth and development of canvasback and lesser scaup ducklings. *Auk* 101: 121–133.

Livezey, B. C. 1996. A phylogenetic analysis of modern pochards. *Wilson Bulletin* 107: 214–234.

Longcore, J. R., and G. W. Cornwell. 1964. The consumption of natural foods by captive canvasbacks and lesser scaups. *Journal of Wildlife Management* 28: 527–531.

Longwell, J. R., and V. Stotts. 1959. Some observations on the recovery of diving ducks banded in the Maryland portion of Chesapeake Bay. *Proceedings Southeastern Association of Game and Fish Commissioners* 12: 285–291.

Mabbott, D. C. 1920. *Food Habits of Seven Species of American Shoal-water Ducks*. Bulletin of the US Department of Agriculture 862. US Department of Agriculture, Washington, DC.

Munro, J. A. 1941. Studies of waterfowl in British Columbia: Greater scaup duck, lesser scaup duck. *Canadian Journal of Research* 19: 113–138.

Nilsson, L. 1970. Food-seeking activity of south Swedish diving ducks in the non-breeding season. *Oikos* 21: 145–154.

Nilsson, L. 1972. Habitat selection, food choice, and feeding habits of diving ducks in coastal waters of south Sweden during the non-breeding season. *Ornis Scandinavica* 3: 55–78.

Noyes, J. H., and R. L. Jarvis. 1985. Diet and nutrition of breeding female redhead and canvasback ducks in Nevada. *Journal of Wildlife Management* 49: 203–211.

Olson, D. P. 1964. A study of canvasback and redhead breeding populations, nesting habitats and productivity. PhD dissertation, University of Minnesota, Minneapolis.

Oring, L. W. 1964. Behavior and ecology of certain ducks during the postbreeding period. *Journal of Wildlife Management* 28: 223–233.

Petrie, S. A., S. S. Badzinski, and K. G. Drouillard. 2007. Contaminants in lesser and greater scaup staging on the lower Great Lakes. *Archives of Environmental Contamination and Toxicology* 52: 580–589.

Reinecker, W. C., and W. Anderson 1960. A waterfowl nesting study on Tule Lake and Klamath National Wildlife Refuges. *California Fish and Game* 46: 481–506.

Ross, R. K., S. A. Petrie, S. S. Badzinski, and A. Mullie. 2005. Autumn diet of greater scaup, lesser scaup, and long-tailed ducks on eastern Lake Ontario prior to zebra mussel invasion. *Wildlife Society Bulletin* 33: 81–91.

Sugden, L. G. 1973. *Feeding Ecology of Pintail, Gadwall, American Widgeon and Lesser Scaup Ducklings in Southern Alberta*. Report Series 24. Canadian Wildlife Service, Ottawa, ON.

Tome, M. W., and D. A. Wrubleski. 1988. Underwater foraging behavior of canvasbacks, lesser scaups, and ruddy ducks. *Condor* 90: 168–172.

Weller, M. W., D. L. Trauger, and G. L. Krapu. 1969. Breeding birds of the West Mirage Islands, Great Slave Lake, NWT. *Canadian Field-Naturalist* 83: 344–360.

Wilson, S. F., and C. D. Ankney. 1988. Variation in structural size and wing stripe of lesser and greater scaup. *Canadian Journal of Zoology* 66: 2045–2048.

Canvasback

Anderson, M. G. 1984. Parental investment and pair-bond behavior among canvasback ducks *(Aythya valisineria*, Anatidae). *Behavioral Ecology and Sociobiology* 15: 81–90.

Anderson, M. G. 1985a. Social behavior of breeding canvasbacks *(Aythya valisineria)*: Male and female strategies of reproduction. PhD dissertation, University of Minnesota, Minneapolis.

Anderson, M. G. 1985b. Variations on monogamy in canvasbacks *(Aythya valisineria)*. *Ornithological Monographs* 37: 57–67.

Anderson, M. G. 1989. Species closures: A case study of the canvasback. *International Waterfowl Symposium* 6: 41–50.

Anderson, M. G., M. S. Lindberg, and R. B. Emery. 2001. Probability of survival and breeding for juvenile female canvasbacks. *Journal of Wildlife Management* 65: 385–397.

Anderson, M. G., R. B. Emery, and T. W. Arnold. 1997. Reproductive success and female survival affect local population density of canvasbacks. *Journal of Wildlife Management* 61: 1174–1191.

Austin, J. E., and J. R. Serie. 1991. Habitat use and movements of canvasback broods in southwestern Manitoba. *Prairie Naturalist* 23: 223–228.

Austin, J. E., and J. R. Serie. 1994. Variation in body mass of wild canvasback and redhead ducklings. *Condor* 96: 909–915.

Austin, J. E., J. R. Serie, and J. N. Noyes. 1990. Diet of canvasbacks during breeding. *Prairie Naturalist* 22: 171–176.

Barzen, J. A., and J. R. Serie. 1989. Patterns of nutrient acquisition in canvasbacks during spring migration. MS thesis, University of North Dakota, Grand Forks.

Barzen, J. A., and J. R. Serie. 1990. Nutrient reserve dynamics of breeding canvasbacks. *Auk* 107: 75–85.

Bluhm, C. K. 1985. Mate preferences and mating patterns of canvasbacks *(Aythya valisineria)*. *Ornithological Monographs* 37: 45–56.

Custer, T. W., and W. L. Hohman. 1994. Trace elements in canvasbacks *(Aythya valisineria)* in Louisiana, USA, 1987–1988. *Environmental Pollution* 84: 253–259.

De Sobrino, C. N. 1995. Brood size, duckling survival and parental care in canvasbacks. MS thesis, University of British Columbia, Vancouver.

Devries, J. H. 1993. Habitat use, movements, and behavior of postbreeding female canvasbacks in Manitoba. MS thesis, Oregon State University, Corvallis.

Doty, H. A., D. L. Trauger, and J. R. Serie. 1984. Renesting by canvasbacks in southwestern Manitoba. *Journal of Wildlife Management* 48: 581–584.

Dzubin, A. 1959. Growth and plumage development of wild-trapped juvenile canvasbacks *(Aythya valisineria)*. *Journal of Wildlife Management* 23: 279–290.

Erickson, R. C. 1948. Life history and ecology of the canvasback, *Nyroca valisineria* (Wilson), in southeastern Oregon. PhD dissertation, Iowa State College, Ames.

Fournier, M. A., and J. E. Hines. 1998. Productivity and population increase of subarctic breeding canvasbacks. *Journal of Wildlife Management* 62: 179–184.

Geis, A. D. 1959. Annual and shooting mortality estimates for the canvasback. *Journal of Wildlife Management* 23: 253–261.

Haramis, G. M. 1991. Canvasback (*Aythya valisineria*). Pp. 17.1–17.10 in S. L. Funderburk, J. A. Mihursky, S. J. Jordan, and D. Riley, eds. *Habitat Requirements for Chesapeake Bay Living Resources.* 2nd ed. Living Resources Subcommittee, Chesapeake Research Consortium, Solomons, MD.

Haramis, G. M., D. G. Jorde, and C. M. Bunck. 1993. Survival of hatching-year canvasbacks wintering on Chesapeake Bay. *Journal of Wildlife Management* 57: 763–771.

Haramis, G. M., D. G. Jorde, S. A. Macko, and J. L. Walker. 2001. Stable-isotope analysis of canvasback winter diet in upper Chesapeake Bay. *Auk* 118: 1,008–1,017.

Haramis, G. M., E. L. Derleth, and W. K. Link. 1994. Flock sizes and sex ratios of canvasbacks in Chesapeake Bay and North Carolina. *Journal of Wildlife Management* 58: 123–131.

Haramis, G. M., J. D. Nichols, K. H. Pollock, and J. E. Hines. 1986. The relationship between body mass and survival of wintering canvasbacks. *Auk* 103: 506–514.

Haramis, G. M., J. R. Goldsberry, D. G. McAuley, and E. L. Derleth. 1985. An aerial photographic census of Chesapeake Bay and North Carolina canvasbacks. *Journal of Wildlife Management* 49: 449–454.

Hochbaum, H. A. 1944. *The Canvasback on a Prairie Marsh.* American Wildlife Institute, Washington, DC.

Hohman, W. L., and D. P. Rave. 1990. Diurnal time-activity budgets of wintering canvasbacks in Louisiana. *Wilson Bulletin* 102: 645–654.

Hohman, W. L., D. W. Woolington, and J. H. Devries. 1990. Food habits of wintering canvasbacks in Louisiana. *Canadian Journal of Zoology* 68: 2,605–2,609.

Hohman, W. L., R. D. Pritchert, J. L. Moore, and D. O. Schaeffer. 1993. Survival of female canvasbacks wintering in coastal Louisiana. *Journal of Wildlife Management* 57: 758–762.

Hohman, W. L., R. D. Pritchert, R. M. Pace, III, D. W. Woolington, and R. Helm. 1990. Influence of ingested lead on body mass of wintering canvasbacks. *Journal of Wildlife Management* 54: 211–215.

Jorde, D. G., G. M. Haramis, C. M. Bunck, and G. W. Pendleton. 1995. Effects of diet on rate of body mass gain by wintering canvasbacks. *Journal of Wildlife Management* 59: 31–39.

Kahl, R. 1991. *Restoration of Canvasback Migrational Staging Habitat in Wisconsin: A Research Plan with Implications for Shallow Lake Management.* Technical Bulletin 172. Wisconsin Department of Natural Resources, Madison.

Korschgen, C. E., K. P. Kenow, W. L. Green, D. H. Johnson, M. D. Samuel, and L. Sileo. 1996. Survival of radiomarked canvasback ducklings in northwestern Minnesota. *Journal of Wildlife Management* 60: 120–132.

Korschgen, C. E., L. S. George, and W. L. Green. 1988. Feeding ecology of canvasbacks staging on Pool 7 of the upper Mississippi River. Pp. 237–249, in M. W. Weller, ed. *Waterfowl in Winter.* University of Minnesota Press, Minneapolis.

Kruse, K. L., J. R. Lovvorn, J. Y. Takekawa, and J. Mackay. 2003. Long-term productivity of canvasbacks (*Aythya valisineria*) in a snowpack-driven desert marsh. *Auk* 120: 107–119.

Leonard, J. P., M. G. Anderson, H. H. Prince, and R. B. Emery. 1996. Survival and movements of canvasback ducklings. *Journal of Wildlife Management* 60: 863–874.

Lightbody, J. P., and C. D. Ankney. 1984. Seasonal influence on the strategies of growth and development of canvasback and lesser scaup ducklings. *Auk* 101: 121–133.

Lovvorn, J. R. 1989. Distributional responses of canvasback ducks to weather and habitat change. *Journal of Applied Ecology* 26: 113–130.

Lovvorn, J. R. 1990. Courtship and aggression in canvasbacks: Influence of sex and pair-bonding. *Condor* 92: 369–378.

Lovvorn, J. R., and J. A. Barzen. 1988. Molt in the annual cycle of canvasbacks. *Auk* 105: 543–552.

Miles, A. K., and H. M. Ohlendorf. 1993. Environmental contaminants in canvasbacks wintering on San Francisco Bay, California. *California Fish and Game* 79: 28–38.

Nichols, J. D., and G. M. Haramis. 1980a. Inferences regarding survival and recovery rates of winter-banded canvasbacks. *Journal of Wildlife Management* 44: 164–173.

Nichols, J. D., and G. M. Haramis. 1980b. Sex-specific differences in winter distribution patterns of canvasbacks. *Condor* 82: 406–416.

Perry, M. C., and F. M. Uhler. 1988. Food habits and distribution of wintering canvasbacks, *Aythya valisineria*, on Chesapeake Bay. *Estuaries* 11: 57–67.

Sayler, R. D. 1996. Behavioral interactions among brood parasites with precocial young: Canvasbacks and redheads on the Delta marsh. *Condor* 98: 801–809.

Serie, J. R., and D. E. Sharp. 1989. Body weight and composition dynamics of fall migrating canvasbacks. *Journal of Wildlife Management* 53: 431–441.

Serie, J. R., D. L. Trauger, and D. E. Sharp. 1983. Migration and winter distributions of canvasbacks staging on the upper Mississippi River. *Journal of Wildlife Management* 47: 741–753.

Serie, J. R., D. L. Trauger, and J. E. Austin. 1992. Influence of age and selected environmental factors on reproductive performance of canvasbacks. *Journal of Wildlife Management* 56: 546–556.

Smith, D. 1946. The canvasback in Minnesota. *Auk* 63: 73–81.

Sorenson, M. D. 1993. Parasitic egg laying in canvasbacks: Frequency, success, and individual behavior. *Auk* 110: 57–69.

Sorenson, M. D. 1997. Effects of intra- and interspecific brood parasitism on a precocial host, the canvasback, *Aythya valisineria*. *Behavioral Ecology* 8: 153–161.

Stewart, R. E., A. D. Geis, and C. D. Evans. 1958. Distribution of populations and hunting kill of the canvasback. *Journal of Wildlife Management* 22: 333–370.

Stoudt, J. H. 1965. *Habitat Requirements of the Canvasback: Progress Report 1965*. Northern Prairie Wildlife Research Center, US Fish and Wildlife Service, Jamestown, ND.

Stoudt, J. H. 1982. *Habitat Use and Productivity of Canvasbacks in Southwestern Manitoba, 1961–72*. Special Scientific Report—Wildlife 248. US Fish and Wildlife Service, Washington, DC.

Sugden, L. G. 1978. *Canvasback Habitat Use and Production in Saskatchewan Parklands*. Occasional Paper 34. Canadian Wildlife Service, Ottawa, ON.

Takekawa, J. Y. 1987. Energetics of canvasbacks staging on an upper Mississippi River pool during fall migration. PhD dissertation, Iowa State University, Ames.

Thompson, J. E. 1992. The nutritional ecology of molting male canvasbacks (*Aythya valisineria*) in central Alberta. MS thesis, University of Missouri–Columbia.

Thompson, J. E., and R. D. Drobney. 1995. Intensity and chronology of postreproductive molts in male canvasbacks. *Wilson Bulletin* 107: 338–358.

Thompson, J. E., and R. D. Drobney. 1996. Nutritional implications of molt in male canvasbacks: Variation in nutrient reserves and digestive tract morphology. *Condor* 98: 512–516.

Thompson, J. E., and R. D. Drobney. 1997. Diet and nutrition of male canvasbacks during postreproductive molts. *Journal of Wildlife Management* 61: 426–434.

White, D. H., R. C. Stendell, and B. M. Mulhern. 1979. Relations of wintering canvasbacks to environmental pollutants, Chesapeake Bay, Maryland. *Wilson Bulletin* 91: 279–287.

Woolington, D. W. 1993. Sex ratios of wintering canvasbacks in Louisiana. *Journal of Wildlife Management* 57: 751–758.

Woolington, D. W., and J. W. Emfinger. 1989. Trends in wintering canvasback populations at Catahoula Lake, Louisiana. *Proceedings of the Southeastern Association of Fish and Wildlife Agencies* 43: 396–403.

Eurasian (Common) Pochard

American Ornithologists' Union. 1998. *Check-list of North American Birds.* 7th ed. Washington, DC.

Dugger, B. D. 1996. The impact of brood parasitism on host fitness in common pochards and tufted ducks. PhD dissertation, University of Missouri, Columbia.

Fox, T. 2005. Common pochard. Pp. 651–654, in J. Kear, ed. *Ducks, Geese and Swans.* Vol. 2. Oxford University Press, Oxford, UK.

Madge, S., and H. Burn. 1988. *Waterfowl: An Identification Guide to the Ducks, Geese and Swans of the World.* Houghton Mifflin, Boston, MA. (Common pochard, pp. 246–248.)

Ogilvie, M. A. 1975. *Ducks of Britain and Europe.* T. & A. D. Poyser, Berkhamsted, UK. (Common pochard, pp. 83–84.)

Owen, M. 1977. *Wildfowl of Europe.* Macmillan, London, UK. (Common pochard, pp. 192–196.)

Patten, M. A. 1993. First record of the common pochard in California. *Western Birds* 24: 235–249.

Petrželková, A., P. Klvaňa, T. Albrecht, and D. Hořák. 2013. Conspecific brood parasitism and host clutch size in common pochards *Aythya ferina. Acta Ornithologica* 48: 103–108.

Pyscgock, J. 2013. #ABArare – Common pochard – Vermont. Lake Champlain, January 2013. American Birding Association blog. http://blog.aba.org/2013/01/abarare-common pochard-vermont.html

Roselaar, C. S. 1977. Pochard. Pp. 561–569, in S. Cramp and K. E. L. Simmons. *Handbook of the Birds of Europe, the Middle East, and North Africa: The Birds of the Western Palearctic, Volume 1, Ostrich to Ducks.* Oxford University Press, Oxford, UK.

Todd, F. S. 1996. *Natural History of the Waterfowl.* Ibis Publishing, Vista, CA. (Eurasian Pochard, pp. 357–358.)

Redhead

Arnold, T. W., M. G. Anderson, M. D. Sorenson, R. B. Emery. 2002. Survival and philopatry of female redheads breeding in southwestern Manitoba. *Journal of Wildlife Management* 66: 162–169.

Austin, J. E., and J. R. Serie. 1994. Variation in body mass of wild canvasback and redhead ducklings. *Condor* 96: 909–915.

Bailey, R. O. 1981. The postbreeding ecology of the redhead duck (*Aythya americana*) on Long Island Bay, Lake Winnipegosis, Manitoba. PhD dissertation, McGill University, Montréal, QC.

Bailey, R. O., and R. D. Titman. 1984. Habitat use and feeding ecology of postbreeding redheads. *Journal of Wildlife Management* 48: 1144–1155.

Ballard, B. M., J. D. James, R. L. Bingham, M. J. Petrie, and B. C. Wilson. 2010. Coastal pond use by redheads wintering in Laguna Madre, Texas. *Wetlands* 30: 669–674.

Benson, D., and L. W. DeGraff. 1968. Distribution and mortality of redheads banded in New York. *New York Fish and Game Journal* 15: 52–70.

Collins, D. P., and R. E. Trost. 2010. *2010 Pacific Flyway Data Book.* Division of Migratory Bird Management, US Fish and Wildlife Service, Portland, Oregon.

Cordts, S. 2010. *2010 Waterfowl Breeding Population Survey, Minnesota.* Minnesota Department of Natural Resources, St. Paul.

Cornelius, S. E. 1977. Food and resource utilization by wintering redheads on lower Laguna Madre. *Journal of Wildlife Management* 41: 374–385.

Cornelius, S. E. 1982. Wetland salinity and salt gland size in the redhead *Aythya americana. Auk* 99: 774–778.

Custer, C. M., T. W. Custer, and P. J. Zwank. 1997. Migration chronology and distribution of redheads on the lower Laguna Madre, Texas. *Southwestern Naturalist* 42: 40–51.

Johnson, D. J. 1978. Age-related breeding biology of the redhead duck in southwestern Manitoba. MS thesis, Texas A&M University, College Station.

Joyner, D. E. 1976. Effects of interspecific nest parasitism by redheads and ruddy ducks. *Journal of Wildlife Management* 40: 33–38.

Joyner, D. E. 1983. Parasitic egg laying in redheads and ruddy ducks in Utah: Incidence and success. *Auk* 100: 717–725.

Kenow, K. P., and D. H. Rusch. 1996. Food habits of redheads at the Horicon Marsh, Wisconsin. *Journal of Field Ornithology* 67: 649–659.

Lightbody, J. P. 1985. Growth rates and development of redhead ducklings. *Wilson Bulletin* 97: 554–559.

Lokemoen, J. T. 1966. Breeding ecology of the redhead duck in western Montana. *Journal of Wildlife Management* 30: 668–681.

Low, J. B. 1945. Ecology and management of the redhead, *Nyroca americana*, in Iowa. *Ecological Monographs* 15: 35–69.

McKnight, D. E. 1974. Dry-land nesting by redheads and ruddy ducks. *Journal of Wildlife Management* 38: 112–119.

Michot, T. C. 2000. Comparison of wintering redhead populations in four Gulf of Mexico seagrass beds. Pp. 243–260 in F. A. Comín, J. A. Herrera, and J. Ramírez, eds. *Limnology and Aquatic Birds: Monitoring, Modeling and Management*. Second International Symposium on Limnology and Aquatic Birds. Universidad Autónoma de Yucatán, Mérida, Yucatán, México.

Michot, T. C., and A. J. Nault. 1993. Diet differences in redheads from nearshore and offshore zones in Louisiana. *Journal of Wildlife Management* 57: 238–244.

Michot, T. C., and P. C. Chadwick. 1994. Potential foods for redheads (*Aythya americana*) in Chandeleur Sound, Louisiana. *Wetlands* 14: 276–283.

Michot, T. C., T. W. Custer, A. J. Nault, and C. M. Mitchell. 1994. Environmental contaminants in redheads wintering in coastal Louisiana and Texas. *Archives of Environmental Contamination and Toxicology* 26: 425–434.

Mitchell, C. A., T. W. Custer, and P. J. Zwank. 1992. Redhead duck behavior on lower Laguna Madre and adjacent ponds of southern Texas. *Southwestern Naturalist* 37: 65–72.

Mitchell, C. A., T. W. Custer, and P. J. Zwank. 1994. Herbivory on shoalgrass by wintering redheads in Texas. *Journal of Wildlife Management* 58: 131–141.

Moore, J. L. 1991. Habitat-related activities and body mass of wintering redhead ducks on coastal ponds in south Texas. MS thesis, Texas A&M University, College Station.

Reinecker, W. C. 1968. A summary of band recoveries from redheads *Aythya americana* banded in northeastern California. *California Fish and Game* 54: 17–26.

Sayler, R. D. 1996. Behavioral interactions among brood parasites with precocial young: Canvasbacks and redheads on the Delta marsh. *Condor* 98: 801–809.

Smart, G. 1965. Development and maturation of primary feathers of redhead ducklings. *Journal of Wildlife Management* 29: 533–536.

Sorenson, M. D. 1991. The functional significance of parasitic egg laying and typical nesting in redhead ducks: An analysis of individual behavior. *Animal Behaviour* 42: 771–796.

Talent, L. G., G. L. Krapu, and R. L. Jarvis. 1981. Effects of redhead nest parasitism on mallards. *Wilson Bulletin* 93: 562–563.

Weller, M. W. 1964. Distribution and migration of the redhead. *Journal of Wildlife Management* 28: 64–103.

Weller, M. W. 1965. Chronology and pair formation in some Nearctic *Aythya* (Anatidae). *Auk* 82: 227–235.

Weller, M. W. 1967. Courtship of the redhead (*Aythya americana*). *Auk* 84: 544–559.

Weller, M. W. 1970. Additional notes on the plumages of the redhead (*Aythya americana*). *Wilson Bulletin* 82: 320–323.

Weller, M. W., and P. Ward. 1959. Migration and mortality of hand-reared redheads (*Aythya americana*). *Journal of Wildlife Management* 23: 427–433.

Williams, S. O., III. 1975. Redhead breeding in the state of Jalisco, México. *Auk* 92: 152–153.

Woodin, M. C. 1996. Wintering ecology of redheads (*Aythya americana*) in the western Gulf of Mexico region. *Gibier Faune Sauvage* 13: 653–665.

Woodin, M. C., and G. A. Swanson. 1989. Foods and dietary strategies of prairie-nesting ruddy ducks and redheads. *Condor* 91: 280–287.

Yerkes, T. 1998. The influence of female age, body mass, and ambient conditions on redhead incubation constancy. *Condor* 100: 62–68.

Yerkes, T. 2000a. Nest-site characteristics and brood-habitat selection of redheads: An association between wetland characteristics and success. *Wetlands* 20: 575–580.

Yerkes, T. 2000b. Influence of female age and body mass on brood and duckling survival, number of surviving ducklings, and brood movements in redheads. *Condor* 102: 926–929.

Ring-necked Duck

Alisauskas, R. T., R. T. Eberhardt, and C. D. Ankney. 1990. Nutrient reserves of breeding ring-necked ducks (*Aythya collaris*). *Canadian Journal of Zoology* 68: 2524–2530.

Conroy, M. J., and R. T. Eberhardt. 1983. Variation in survival and recovery rates of ring-necked ducks. *Journal of Wildlife Management* 47: 127–137.

Eberhardt, R. T., and M. Riggs. 1995. Effects of sex and reproductive status on the diets of breeding ring-necked ducks (*Aythya collaris*) in north-central Minnesota. *Canadian Journal of Zoology* 73: 392–399.

Erskine, A. J. 1972. Postbreeding assemblies of ring-necked ducks in eastern Nova Scotia. *Auk* 89: 449–450.

Evrard, J. O., B. R. Bacon, and T. R. Grunewald. 1987. Unusual upland nests of the ring-necked duck. *Journal of Field Ornithology* 58: 31.

Hohman, W. L. 1984a. Aspects of the breeding biology of ring-necked ducks (*Aythya collaris*). PhD dissertation, University of Minnesota–St. Paul, St. Paul, MN.

Hohman, W. L. 1984b. Diurnal time-activity budgets for ring-necked ducks wintering in central Florida. *Proceedings of the Southeastern Association of Fish and Wildlife Agencies* 38: 158–164.

Hohman, W. L. 1985. Feeding ecology of breeding ring-necked ducks in northwestern Minnesota. *Journal of Wildlife Management* 49: 546–557.

Hohman, W. L. 1986a. Incubation rhythms of ring-necked ducks. *Condor* 88: 290–296.

Hohman, W. L. 1986b. Changes in body weight and body composition of breeding ring-necked ducks (*Aythya collaris*). *Auk* 103: 181–188.

Hohman, W. L., and M. W. Weller. 1994. Body mass and composition of ring-necked ducks wintering in southern Florida. *Wilson Bulletin* 106: 494–507.

Hohman, W. L., and R. D. Crawford. 1995. Molt in the annual cycle of ring-necked ducks. *Condor* 97: 473–483.

Hohman, W. L., T. S. Taylor, and M. W. Weller. 1988. Annual body weight change in ring-necked ducks (*Aythya collaris*). Pp. 257–269, in M. W. Weller, ed. *Waterfowl in Winter*. University of Minnesota Press, Minneapolis.

Jeske, C. W., and H. F. Percival. 1995. Time and energy budgets of wintering ring-necked ducks *Aythya collaris* in Florida, USA. *Wildfowl* 46: 109–118.

Jeske, C. W., H. F. Percival, and J. E. Thul. 1995. Food habits of ring-necked ducks wintering in Florida. *Proceedings of the Southeastern Association of Fish and Wildlife Agencies* 47: 130–137.

Koons, D. N., and J. J. Rotella. 2003. Comparative nesting success of sympatric lesser scaup and ring-necked ducks. *Journal of Field Ornithology* 74: 222–229.

Maxson, S. J., and R. M. Pace, III. 1992. Diurnal time-activity budgets and habitat use of ring-necked duck ducklings in north-central Minnesota. *Wilson Bulletin* 104: 472–484.

McAuley, D. G., and J. R. Longcore. 1988a. Survival of juvenile ring-necked ducks: Relationship to wetland pH. *Journal of Wildlife Management* 52: 169–176.

McAuley, D. G., and J. R. Longcore. 1988b. Foods of juvenile ring-necked ducks: Relationship to wetland pH. *Journal of Wildlife Management* 52: 177–185.

McAuley, D. G., and J. R. Longcore. 1989. Nesting phenology and success of ring-necked ducks in east-central Maine. *Journal of Field Ornithology* 60: 112–119.

Mendall, H. L. 1958. *The Ring-necked Duck in the Northeast*. University of Maine Studies, Second Series, Number 73. University of Maine Press, Orono.

Peters, M. S., and A. D. Afton. 1993a. Diets of ring-necked ducks wintering on Catahoula Lake, Louisiana. *Southwestern Naturalist* 38: 166–168.

Peters, M. S., and A. D. Afton. 1993b. Blood lead concentrations and ingested shot in ring-necked ducks at Catahoula Lake, Louisiana. *Proceedings of the Southeastern Association of Fish and Wildlife Agencies* 47: 292–298.

Ripley, S. D. 1963. Courtship in the ring-necked duck. *Wilson Bulletin* 75: 373–375.

Stathis, N. A. 1994. Fall migration ecology of ring-necked ducks in east-central Minnesota. MS thesis, University of Minnesota–St. Paul, St. Paul, MN.

Tufted Duck

Austin, G. T. 1969. A record of the tufted duck for Connecticut. *Wilson Bulletin* 81: 332.

Bengtson, S.-A. 1970. Location of nest-sites of ducks in Lake Mývatn area, north-east Iceland. *Oikos* 21: 218–229.

Dugger, B. D. 1996. The impact of brood parasitism on host fitness in common pochards and tufted ducks. PhD dissertation, University of Missouri, Columbia.

Goochfield, M. 1968. Notes on the status of the tufted duck (*Aythya fuligula*) in North America with a report of a new observation from Wyoming. *Condor* 70: 186–187.

Hildén, O. 1964. Ecology of duck populations in the islands group of Valassaaret, Gulf of Bothnia. *Annales Zoologici Fennici* 1: 153–279.

Johnsgard, P. A. 1978. *Ducks, Geese and Swans of the World.* University of Nebraska Press, Lincoln. (Tufted duck, pp. 300–303.)

Kear, J. 1970. Studies on the development of young tufted duck. *Wildfowl* 21: 123–132.

Madge, S., and H. Burn. 1988. *Waterfowl: An Identification Guide to the Ducks, Geese and Swans of the World.* Houghton Mifflin, Boston. (Tufted duck, pp. 255–256.)

Ogilvie, M. A. 1975. *Ducks of Britain and Europe.* T. & A. D. Poyser, Berkhamsted, UK. (Tufted duck, pp. 300–303.)

Olney, P. S. J. 1965. The food and feeding habits of the tufted duck, *Aythya fuligula. Ibis* 105: 55–62.

Owen, M. 1977. *Wildfowl of Europe.* Macmillan, London, UK. (Tufted duck, pp. 198–201.)

Roselaar, C. S. 1977. Tufted duck. Pp. 577–586, in S. Cramp and K. E. L. Simmons. *Handbook of the Birds of Europe, the Middle East, and North Africa: The Birds of the Western Palearctic, Volume 1, Ostrich to Ducks.* Oxford University Press, Oxford, UK.

Todd, F. S. 1996. *Natural History of the Waterfowl.* Ibis Publishing, Vista, CA. (Tufted duck, pp. 369–370.)

Greater Scaup

Banks, R. C. 1986. Subspecies of the greater scaup and their names. *Wilson Bulletin* 98: 433–444.

Billard, R. S., and P. S. Humphrey. 1972. Molts and plumages in the greater scaup. *Journal of Wildlife Management* 36: 765–774.

Burger, J. 1983. Jamaica Bay studies, VI: Factors affecting distribution of greater scaup *Aythya marila* in a coastal estuary in New York, USA. *Ornis Scandinavica* 14: 309–316.

Cohen, J. B. 1998. Greater scaup as bioindicators of contaminants in Long Island Sound. MS thesis, University of Connecticut, Storrs.

Cohen, J. B., J. S. Barclay, A. R. Major, and J. P. Fisher. 2000. Wintering greater scaup as biomonitors of metal contamination in federal wildlife refuges in the Long Island region. *Archives of Environmental Contamination and Toxicology* 38: 83–92.

Flint, P. L. 2003. Incubation behaviour of greater scaup *Aythya marila* on the Yukon-Kuskokwim delta, Alaska. *Wildfowl* 54: 71–79.

Flint, P. L., and J. B. Grand. 1999. Patterns of variation in size and composition of greater scaup eggs: Are they related? *Wilson Bulletin* 111: 465–471.

Flint, P. L., J. B. Grand, T. F. Fondell, and J. A. Morse. 2006. *Population Dynamics of Greater Scaup Breeding on the Yukon-Kuskokwim Delta, Alaska.* Wildlife Monographs 162. The Wildlife Society and John Wiley & Sons.

Longley, W. H. 1949. Greater scaups eating frogs. *Auk* 66: 200.

McAlpine, D. F., S. Makepeace, and M. Phinney. 1988. Breeding records of the greater scaup, *Aythya marila*, in New Brunswick. *Canadian Field-Naturalist* 102: 718–719.

Petrie, S. A., S. S. Badzinski, and K. G. Drouillard. 2007. Contaminants in lesser and greater scaup staging on the lower Great Lakes. *Archives of Environmental Contamination and Toxicology* 52: 580–589.

Rocque, D. A. 1997. Population ecology and modeling of greater scaup. MS thesis, University of Connecticut, Storrs.

Vermeer, K., D. R. M. Hatch, and J. A. Windsor. 1972. Greater scaup is common breeder on northern Lake Winnipeg. *Canadian Field-Naturalist* 86: 168.

Wahle, L. C., and J. S. Barclay. 1993. Changes in greater scaup foods in Connecticut. *Northeast Wildlife* 50: 69–76.

Ware, L. L., S. A. Petrie, S. S. Badzinski, and R. C. Bailey. 2010. Selenium concentrations in greater scaup and dreissenid mussels during winter on western Lake Ontario. *Archives of Environmental Contamination and Toxicology* 61: 292–299.

Lesser Scaup

Afton, A. D. 1983. Male and female strategies for reproduction in lesser scaup. PhD dissertation, University of North Dakota, Grand Forks.

Afton, A. D. 1984. Influence of age and time on reproductive performance of female lesser scaup. *Auk* 101: 255–265.

Afton, A. D. 1985. Forced copulation as a reproductive strategy of male lesser scaup: A field test of some predictions. *Behaviour* 92: 146–167.

Afton, A. D. 1993. Post-hatch brood amalgamation in lesser scaup: Female behavior and return rates, and duckling survival. *Prairie Naturalist* 25: 227–235.

Afton, A. D., and C. D. Ankney. 1991. Nutrient-reserve dynamics of breeding lesser scaup: A test of competing hypotheses. *Condor* 93: 89–97.

Afton, A. D., and M. G. Anderson. 2001. Declining scaup populations: A retrospective analysis of long-term population and harvest survey data. *Journal of Wildlife Management* 65: 781–796.

Afton, A. D., R. H. Hier, and S. L. Paulus. 1991. Lesser scaup diets during migration and winter in the Mississippi Flyway. *Canadian Journal of Zoology* 69: 328–333.

Anteau, M. J., and A. D. Afton. 2006. Diet shifts of lesser scaup are consistent with the spring condition hypothesis. *Canadian Journal of Zoology* 84: 779–786.

Anteau, M. J., and A. D. Afton. 2008. Diets of lesser scaup during spring migration throughout the upper Midwest are consistent with the spring condition hypothesis. *Waterbirds* 31: 97–106.

Anteau, M. J., and A. D. Afton. 2009. Wetland use and feeding by lesser scaup during spring migration across the upper Midwest, USA. *Wetlands* 29: 704–712.

Anteau, M. J., A. D. Afton, C. M. Custer, and T. W. Custer. 2007. Relationships of cadmium, mercury, and selenium with nutrient reserves of female lesser scaup (*Aythya affinis*) during winter and spring migration. *Environmental Toxicology and Chemistry* 26: 515–520.

Austin, J. E. 1987. Activities of postbreeding lesser scaup in southwestern Manitoba. *Wilson Bulletin* 99: 448–456.

Austin, J. E., and L. H. Fredrickson. 1986. Molt of female lesser scaup immediately following breeding. *Auk* 103: 293–298.

Austin, J. E., and L. H. Fredrickson. 1987. Body and organ mass and body composition of postbreeding female lesser scaup. *Auk* 104: 694–699.

Bartonek, J. C., and H. W. Murdy. 1970. Summer foods of lesser scaup in subarctic taiga. *Arctic* 23: 35–44.

Brady, C. M. 2009. Effects of dietary selenium on the health and survival of wintering lesser scaup. MS thesis, University of Western Ontario, London, ON.

Brook, R. W., and R. G. Clark. 2005. Breeding season survival of female lesser scaup in the northern boreal forest. *Arctic* 58: 16–20.

Chabreck, R. H., and T. Takagi. 1985. Foods of lesser scaup in crayfish impoundments in Louisiana. *Proceedings of the Southeastern Association of Fish and Wildlife Agencies* 39: 465–470.

Corcoran, R. M., J. R. Lovvorn, M. R. Bertram, and M. T. Vivion. 2007. Lesser scaup nest success and duckling survival on the Yukon Flats, Alaska. *Journal of Wildlife Management* 71: 127–134.

Custer, C. M., T. W. Custer, M. J. Anteau, A. D. Afton, and D. E. Wooton. 2003. Trace elements in lesser scaup (*Aythya affinis*) from the Mississippi Flyway. *Ecotoxicology* 12: 47–54.

Dawson, R. D., and R. G. Clark. 1996. Effects of variation in egg size and hatching date on survival of lesser scaup *Aythya affinis* ducklings. *Ibis* 138: 693–699.

Dirschl, H. J. 1969. Foods of lesser scaup and blue-winged teal in the Saskatchewan River delta. *Journal of Wildlife Management* 33: 77–87.

Esler, D., J. B. Grand, and A. D. Afton. 2001. Intraspecific variation in nutrient reserve use during clutch formation by lesser scaup. *Condor* 103: 810–820.

Fast, P. L. F., R. G. Clark, R. W. Brook, and J. E. Hines. 2004. Patterns of wetland use by brood-rearing lesser scaup in the northern boreal forest of Canada. *Waterbirds* 27: 177–182.

Fox, G. A., M. C. MacCluskie, and R. W. Brook. 2005. Are current contaminant concentrations in eggs and breeding female lesser scaup of concern? *Condor* 107: 50–61.

Gehrman, K. H. 1951. An ecological study of the lesser scaup duck (*Aythya affinis* Eyton) at West Medical Lake, Spokane County, Washington. MS thesis, State College of Washington, Pullman, WA.

Hammell, G. S. 1973. The ecology of the lesser scaup (*Aythya affinis* Eyton) in southwestern Manitoba. MS thesis, University of Guelph, ON.

Harmon, B. G. 1962. Mollusks as food of lesser scaup along the Louisiana coast. *North American Wildlife and Natural Resources Conference Transactions* 27: 132–138.

Herring, G., and J. A. Collazo. 2004. Winter survival of lesser scaup in east-central Florida. *Journal of Wildlife Management* 68: 1082–1087.

Hines, J. E. 1977. Nesting and brood ecology of lesser scaup at Waterhen Marsh, Saskatchewan. *Canadian Field-Naturalist* 91: 248–255.

Koons, D. N. 2000. First record of brown-headed cowbird egg in a lesser scaup nest. *Wilson Bulletin* 112: 554.

McKnight, D. E., and I. O. Buss. 1962. Evidence of breeding in yearling female lesser scaup. *Journal of Wildlife Management* 26: 328–329.

Pace, R. M., III, and A. D. Afton. 1999. Direct recovery rates of lesser scaup banded in northwest Minnesota: Sources of heterogeneity. *Journal of Wildlife Management* 63: 389–395.

Pilatzki, A. E., R. D. Neiger, S. R. Chipps, K. F. Higgins, N. Thiex, and A. D. Afton. 2011. Hepatic element concentrations of lesser scaup (*Aythya affinis*) during spring migration in the upper Midwest. *Archives of Environmental Contamination and Toxicology* 61: 144–150.

Richman, S. E., and J. R. Lovvorn. 2004. Relative foraging value to lesser scaup ducks of native and exotic clams from San Francisco Bay. *Ecological Applications* 14: 1217–1231.

Rogers, J. P. 1959. Low water and lesser scaup reproduction near Erickson, Manitoba. *North American Wildlife Conference Transactions* 24: 216–224.

Rogers, J. P. 1964. Effect of drought on reproduction of the lesser scaup. *Journal of Wildlife Management* 28: 213–222.

Rogers, J. P., and L. J. Korschgen. 1966. Foods of lesser scaups on breeding, migration, and wintering areas. *Journal of Wildlife Management* 30: 258–264.

Rotella, J. J., R. G. Clark, and A. D. Afton. 2003. Survival of female lesser scaup: Effects of body size, age, and reproductive effort. *Condor* 105: 336–347.

Siegfried, W. R. 1974. Time budget of behavior among lesser scaups on Delta Marsh. *Journal of Wildlife Management* 38: 708–713.

Smith, R. I. 1963. *Lesser Scaup and Ring-necked Duck Shooting Pressure and Mortality Rates*. Administrative Report No. 20, US Fish and Wildlife Service, Washington, DC.

Stephenson, R. 1994. Diving energetics in lesser scaup (*Aythya affinis* Eyton). *Journal of Experimental Biology* 190: 155–178.

Strand, K. A. 2005. Diet and body composition of migrating lesser scaup (*Aythya affinis*) in eastern South Dakota. MS thesis, South Dakota State University, Brookings.

Strand, K. A., S. R. Chipps, S. N. Kahara, K. F. Higgins, and S. Vaa. 2008. Patterns of prey use by lesser scaup *Aythya affinis* (Aves) and diet overlap with fishes during spring migration. *Hydrobiologia* 598: 389–398.

Tebbs, R. J. 1995. Food habits of lesser scaup, *Aythya affinis*, on lower Pool 19, Mississippi River. MS thesis, Western Illinois University, Macomb.

Trauger, D. L. 1971. Population ecology of lesser scaup (*Aythya affinis*) in subarctic taiga. PhD dissertation, Iowa State University, Ames.

Trauger, D. L. 1974. Eye color of female lesser scaup in relation to age. *Auk* 91: 243–254.

Turnbull, R. E., D. H. Brakhage, and F. A. Johnson. 1986. Lesser scaup mortality from commercial trotlines on Lake Okeechobee, Florida. *Proceedings of the Southeastern Association of Fish and Wildlife Agencies* 40: 465–469.

Stiff-tailed Ducks: Multiple and World Taxa

Blake, E. R. 1977. *Manual of Neotropical Birds.* Vol. 1. University of Chicago Press, Chicago, IL. (Masked duck, pp. 257–258; ruddy duck pp. 256–257.)

Callaghan, D., J. Kear, and K. McCracken. 2005. The stiff-tailed ducks and their allies. Pp. 343–346, in J. Kear, ed. *Ducks, Geese and Swans.* Vol. 1. Oxford University Press, Oxford, UK.

Carbonell, M. 1983, Comparative studies of stiff-tailed ducks (Tribe Oxyurini, Family Anatidae). PhD dissertation, University College, Cardiff, Wales.

Gomez-Dallmeier, F., and A. Cringan. 1990. *Biology, Conservation and Management of Waterfowl in Venezuela.* Editorial ex Libris, Caracas, Venezuela.

Guay, P.-J., and R. A. Mulder. 2007. Skewed paternity distribution in the extremely size dimorphic musk duck (*Biziura lobata*). *Emu* 107: 3, 190.

Hill, S. L, and W. L. Brown. 1986. *A Guide to the Birds of Colombia.* Princeton University Press, Princeton, NJ. (Masked duck, pp. 86–87; ruddy duck, p. 86.)

Howell, S. N. G., and S. Webb. 1995. *A Guide to the Birds of Mexico and Northern Central America.* Oxford University Press, Oxford, UK. (Masked duck, pp. 172–173; ruddy duck, p. 172.)

Johnsgard, P. A. 1965. Observations on some aberrant Australian Anatidae. *Wildfowl Trust 16th Annual Report*, pp. 73–83.

Johnsgard, P. A. 1966. Behavior of the Australian musk duck and blue-billed duck. *Auk* 83: 98–110. http://digitalcommons.unl.edu/biosciornithology/61

Johnsgard, P. A. 1968. Some observations on maccoa duck behavior. *Ostrich* 39: 219–222. http://digitalcommons.unl.edu/johnsgard/15

Johnsgard, P. A., and M. Carbonell. 1996. *Ruddy Ducks and Other Stifftails: Their Behavior and Biology.* University of Oklahoma Press, Norman. (Masked duck, pp. 130–141; ruddy duck, pp. 142–175.)

Leopold, S. 1959. *Wildlife of Mexico: The Game Birds and Mammals.* University of California Press, Berkeley.

Livezey, B. 1995. Phylogeny and comparative ecology of the stiff-tailed ducks (Anatidae: Oxyurini). *Wilson Bulletin* 107: 214–234.

Lockwood, M. W., and B. Freeman. 2014. *Handbook of Texas Birds.* 2nd ed. Texas A&M University Press, College Station. (Masked duck, pp. 36–37; ruddy duck, p. 37.)

McCracken, K. G. 2000. The 20 cm spiny penis of the Andean lake duck (*Oxyura vittata*). *Auk* 117: 820–825.

McCracken, K. G., and M. D. Sorenson. 2005. Is homoplasy or lineage sorting the source of incongruent mDNA and nuclear gene trees in the stuff-tailed ducks (*Nomonyx–Oxyura*)? *Systematic Biology* 54: 35–55.

Oberholser, H. C. 1974. *The Bird Life of Texas.* Vol. 1. University of Texas Press, Austin. (Masked duck, pp. 194–195; ruddy duck, pp. 192–194.)

Owen, M., and S. Young. 1998. *Wildfowl of the World.* New Holland Publishing, Cape Town, South Africa. (Masked duck, pp. 1967; ruddy duck, pp. 168.)

Ridgely, R. S., and J. A Gwyne. 1989. *A Guide to the Birds of Panama.* Princeton University Press, Princeton, NJ. (Masked duck, p. 82.)

Ridgely, R. S, and P. J. Greenfield. 2001. *The Birds of Ecuador.* Cornell University Press, Ithaca, NY. (Masked duck, p. 69.)

Robertson, W. B., Jr., and G. E. Woolfenden. 1992. *Florida Bird Species—An Annotated List.* Florida Ornithological Society, Special Publication No. 6, Gainesville, FL.

Sick, H. 1993. *Birds in Brazil: A Natural History.* Princeton University Press, Princeton, NJ. (Masked duck, pp. 163–164.)

Stiles, F. G., and A. F. Skutch. 1989. *A Guide to the Birds of Costa Rica.* Cornell University Press, Ithaca, NY. (Masked duck, p. 95.)

Todd, F. S. 1996. *Natural History of the Waterfowl.* Ibis Publishing, Vista, CA. (Masked duck, pp. 447–450; ruddy duck, pp. 435–440.)

Wetmore, A. 1965. *The Birds of the Republic of Panama, Part 1.* Smithsonian Miscellaneous Collections 150. Smithsonian Institution, Washington, DC. (Masked duck, pp. 150–153.)

Masked Duck

Anderson, J. T., and T. C. Tacha. 1999. Habitat use by masked ducks along the Gulf coast of Texas. *Wilson Bulletin* 111: 119–121.

Anderson, J. T., G. T. Muehl, and T. C. Tacha. 1998. Distribution and abundance of waterbirds in coastal Texas. *Bird Populations* 4: 1–15.

Anderson, J. T., G. T. Muehl, T. C. Tacha, and D. S. Lobpries. 2000. Wetland use by non-breeding ducks in coastal Texas, USA. *Wildfowl* 51: 191–214.

Barbour, T. 1923. *The Birds of Cuba*. Memoirs of the Nuttall Ornithological Club 6. Nuttall Ornithological Club, Cambridge, MA.

Berrett, D. G. 1962. The birds of the Mexican state of Tabasco. PhD dissertation, Louisiana State University, Baton Rouge.

Blankenship, T. L., and J. T. Anderson. 1993. A large concentration of masked ducks (*Oxyura dominica*) on the Welder Wildlife Refuge, San Patricio County, Texas. *Bulletin of the Texas Ornithological Society* 26: 19–21.

Bond, J. 1961. *Sixth Supplement to the Check-list of Birds of the West Indies* (1956). Academy of Natural Sciences, Philadelphia, PA. 12 pp.

Bowman, M. C. 1995. Sighting of masked duck ducklings in Florida. *Florida Field Naturalist* 23: 35.

Delnicki, D. 1975. The masked duck. *Ducks Unlimited* 41: 46–60.

Eitniear, J. 2010. Noteworthy breeding of masked ducks in Live Oak County, Texas. *Bulletin of the Texas Ornithological Society* 43(1–2): 87–88.

Eitniear, J., and K. Rylander. 2008. Mandibular structure in ruddy and masked ducks: Does morphology reflect ecology? *Texas Journal of Science* 60: 49.

Eitniear, J., and M. J. Morel. 2012. A large concentration of masked ducks in Puerto Rico. *Journal of Caribbean Ornithology* 28: 92–94.

Eitniear, J. C. 2000. Masked duck. In *The Texas Breeding Bird Atlas*. http://txtbba.tamu.edu/species-accounts/masked-duck/

Eitniear, J. C., and S. Colón. 2005. Recent observations of masked duck (*Nomonyx dominica*) in the Caribbean. *Caribbean Journal of Science* 41: 861–864.

Hughes, B. 2005. Masked duck. Pp. 348–351, in J. Kear, ed. *Ducks, Geese and Swans*. Vol. 1. Oxford University Press, Oxford, UK.

Jenni, D. A. 1969. Diving times of the least grebe and masked duck. *Auk* 86: 355–356.

Johnsgard, P. A., and D. Hagemeyer. 1969. The masked duck in the United States. *Auk* 86: 691–695.

Lockwood, M. W. 1997. A closer look: Masked duck. *Birding* 29: 386–390.

Ruddy Duck

Alisauskas, R. T., and C. D. Ankney. 1994a. Costs and rates of egg formation in ruddy ducks. *Condor* 96: 11–18.

Alisauskas, R. T., and C. D. Ankney. 1994b. Nutrition of breeding female ruddy ducks: The role of nutrient reserves. *Condor* 96: 878–897.

Bennett, J. L. 1938. Redheads and ruddy ducks nesting in Iowa. *North American Wildlife Conference Transactions* 3: 647–650.

Boon, L. A., and C. D. Ankney. 1999. Body size, nest initiation date, and egg production in ruddy ducks. *Auk* 116: 228–231.

Brua, R. B. 1997. Brood survival and movements of ruddy ducks in southwestern Manitoba [abstract]. *North American Duck Symposium and Workshop* 1: 33–34.

Brua, R. B. 1998. Factors affecting reproductive success of male and female ruddy ducks (*Oxyura jamaicensis*) in south-eastern Manitoba. PhD dissertation, University of Dayton, Dayton, OH.

Brua, R. B. 1999. Ruddy duck nesting success: Do nest characteristics deter nest predation? *Condor* 101: 867–870.

Brua, R. B., and K. L. Machin. 2000. Determining and testing the accuracy of incubation stage of ruddy duck eggs by floatation. *Wildfowl* 51: 181–189.

Euliss, N. H., Jr., R. L. Jarvis, and D. S. Gilmer. 1989. Carbonate deposition on tail feathers of ruddy ducks using evaporation ponds. *Condor* 91: 803–806.

Gray, B. J. 1980. Reproduction, energetics, and social structure of the ruddy duck. PhD dissertation, University of California–Davis.

Hays, H., and H. M. Habermann. 1969. Note on bill color of the ruddy duck, *Oxyura jamaicensis rubida*. *Auk* 86: 765–766.

Henderson I. S. 2010. The eradication of ruddy ducks in the United Kingdom. *Aliens: The Invasive Species Bulletin* 29: 17–24.

Hobson, K. A., R. B. Brua, W. L. Hohman, and L. I. Wassenaar. 2000. Low frequency of "double molt" of remiges in ruddy ducks revealed by stable isotopes: Implications for tracking migratory waterfowl. *Auk* 117: 129–135.

Hohman, W. L. 1993. Body composition dynamics of ruddy ducks during wing moult. *Canadian Journal of Zoology* 71: 2224–2228.

Hohman, W. L. 1996. Prevalence of double wing molt in free-living ruddy ducks. *Southwestern Naturalist* 41: 195–198.

Hughes, B. 1990. The ecology and behaviour of the North American ruddy duck *Oxyura jamaicensis* in Great Britain and its interaction with native waterbirds: A progress report. *Wildfowl* 41: 133–138.

Hughes, B. 2005. Ruddy duck. Pp. 351–355, in J. Kear, ed. *Ducks, Geese and Swans*. Vol. 1. Oxford University Press, Oxford, UK.

Jehl, J. R., Jr., and E. Johnson. 2004. Wing and tail molts of the ruddy duck. *Waterbirds* 27: 54–59.

Johnsgard, P. A. 1956. Effects of water fluctuation and vegetation change on bird populations, especially waterfowl. *Ecology* 37: 689–701.

Johnsgard, P. A., and C. Nordeen, 1981. Observations on the displays and relationships of the Argentine blue-billed duck (*Oxyura vittata*). *Wildfowl* 32: 5–9.

Joyner, D. E. 1969. A survey of the ecology and behavior of the ruddy duck (*Oxyura jamaicensis*) in northern Utah. MS thesis, University of Utah, Salt Lake City.

Joyner, D. E. 1975. Nest parasitism and brood-related behavior of the ruddy duck. PhD dissertation, University of Nebraska–Lincoln.

Joyner, D. E. 1976. Effects of interspecific nest parasitism by redheads and ruddy ducks. *Journal of Wildlife Management* 40: 33–38.

Joyner, D. E. 1977a. Nest desertion by ruddy ducks in Utah. *Bird-Banding* 48: 19–24.

Joyner, D. E. 1977b. Behavior of ruddy duck broods in Utah. *Auk* 94: 343–349.

Joyner, D. E. 1983. Parasitic egg laying in redheads and ruddy ducks in Utah: Incidence and success. *Auk* 100: 717–725.

Libby, H. J. 1972. Ruddy duck distribution in relation to marsh habitat. MS thesis, University of Wisconsin, Madison.

Low, J. B. 1941. Nesting of the ruddy duck in Iowa. *Auk* 58: 506–517.

McKnight, D. E. 1974. Dry-land nesting by redheads and ruddy ducks. *Journal of Wildlife Management* 38: 112–119.

Miller, M. R., R. M. McLandress, and B. J. Gray. 1977. The display flight of the North American ruddy duck. *Auk* 94: 140–142.

Misterek, D. L. 1974. The breeding ecology of the ruddy duck (*Oxyura jamaicensis*) on Rush Lake, Winnebago County, Wisconsin. MS thesis, University of Wisconsin–Oshkosh.

Muñoz-Fuentes, V., C. Vilà, A. J. Green, J. J. Negro, and M. D. Sorenson. 2007. Hybridization between white-headed ducks and introduced ruddy ducks in Spain. *Molecular Ecology* 16: 629–638.

Pelayo, J. T. 2001. Correlates and consequences of egg size variation in wild ruddy ducks (*Oxyura jamaicensis*). MS thesis, University of Saskatchewan, Saskatoon.

Pelayo, J. T., and R. G. Clark. 2002. Variation in size, composition, and quality of ruddy duck eggs and ducklings. *Condor* 104: 457–462.

Reichart, L. M., S. Anderholm, V. Muñoz-Fuentes, and M. S. Webster. 2010. Molecular identification of brood-parasitic females reveals an opportunistic reproductive tactic in ruddy ducks. *Molecular Ecology* 19: 401–413.

Rienecker, W. C. 1968, and W. Anderson 1960. A waterfowl nesting study on Tule Lake and Klamath National Wildlife Refuges. *California Fish and Game* 46: 481–506.

Ripley, S. D., and G. E. Watson. 1956. Cuban bird notes. *Postilla*, 26: 1–6.

Rohwer, F. C., and S. Freeman. 1989. The distribution of conspecific nest parasitism in birds. *Canadian Journal of Zoology* 67: 239–253.

Siegfried, W. R. 1973a. Post-embryonic development of the ruddy duck *Oxyura jamaicensis* and some other diving ducks. *International Zoo Yearbook* 13: 77–87.

Siegfried, W. R. 1973b. Summer food and feeding of the ruddy duck in Manitoba. *Canadian Journal of Zoology* 51: 1293–1297.

Siegfried, W. R. 1976a. Breeding biology and parasitism in the ruddy duck. *Wilson Bulletin* 88: 566–574.

Siegfried, W. R. 1976b. Segregation in feeding behaviour of four diving ducks in southern Manitoba. *Canadian Journal of Zoology* 54: 730–736.

Siegfried, W. R. 1977. Notes on behaviour of ruddy ducks during the brood period. *Wildfowl* 28: 126–128.

Siegfried, W. R., A. E. Burger, and P. J. Caldwell. 1976. Incubation behavior of ruddy and maccoa ducks. *Condor* 78: 512–517.

Somerville, A. J. 1985. Advantages to late breeding in ruddy ducks. MS thesis, University of British Columbia, Vancouver.

Stark, R. T. 1978. Food habits of the ruddy duck (*Oxyura jamaicensis*) at the Tinicum National Environmental Center. MS thesis, Penn State University, State College.

Tome, M. W. 1987. An observation of renesting by a ruddy duck, *Oxyura jamaicensis*. *Canadian Field-Naturalist* 101: 153–154.

Tome, M. W. 1989. Search-path characteristics of foraging ruddy ducks. *Auk* 106: 42–48.

Tome, M. W. 1991. Diurnal activity budget of female ruddy ducks breeding in Manitoba. *Wilson Bulletin* 103: 183–189.

Wetmore, A. 1917. On certain secondary characteristics of the male ruddy duck, *Erismatura jamaicensis*. *Proceedings US National Museum* 219: 479–482.

Wetmore, A. 1918. A note on the tracheal air sac of the ruddy duck. *Condor* 20: 19–20.

White, D. H., and T. E. Kaiser. 1976. Residues of organochlorines and heavy metals in ruddy ducks from the Delaware River, 1973. *Pesticide Monitoring Journal* 9: 155–156.

Woodin, M. C., and G. A. Swanson. 1989. Foods and dietary strategies of prairie-nesting ruddy ducks and redheads. *Condor* 91: 280–287.

www.ingramcontent.com/pod-product-compliance
Lightning Source LLC
Chambersburg PA
CBHW080331270326
41927CB00014B/3178